Advance praise for *Farm the City*

Most of the world's people live in cities, and *Farm the City* is a story of how to bring cities back to life, literally and emotionally. The cold, forbidding landscapes of urban life bring our hearts to a standstill. When streets, medians, abandoned land, parks, and byways are transformed by soil, bugs, microbes, pollinators, and seeds, lives bloom, connectedness flourishes, and people become denizens once again. Local food is not a mere talisman or gesture. We localize food webs near our homes for identity, nourishment, and taste. Taste is a sense, but it is also a common sense. Local food not only addresses quality of life, economy, and food security, it changes our hearts. Michael Ableman has a finely honed sensibility. Read how he gardens society, grows well-being, weeds out despair, and sows hope in this wonderfully written testament to life.

— Paul Hawken, author, *Drawdown* and *Blessed Unrest*

Michael Ableman documents that generating paradise by growing vegetables amidst the urban jungle also rehabilitates lost souls, builds community, and creates genuine economic value.

— Gabor Maté, MD, author, *In the Realm of Hungry Ghosts*

There is much to admire and emulate in the work of Sole Food Street Farms. Their commitment to create employment for people with barriers, ingenuity in container orcharding, and diversified marketing schemes are just a few examples of their profoundly practical and ethical approach to urban farming. Ableman's *Farm the City* is an inspiring how-to guide for any urban grower who is serious about success.

— Leah Penniman, co-director, Soul Fire Farm,
author, *Farming While Black*

Michael Ableman is one of the handful of inspiring visionaries on the planet who are redefining our future food systems.

— Patrick Holden, founding director, Sustainable Food Trust

So you think you want to farm in the city? In this book, Ableman raises that question. It is more than a toolkit on growing food. He explains what it takes to be a farmer in municipalities daunted with rules and regulations. Yet in the end he explains that farming is more than a passion, it's a business. This book is a testament to all farmers: that our hard work has value and the food we grow and sell for any reason is invaluable.

— Karen Washington, farmer-activist, Rise and Root Farm

This man really connects with his readers. You can feel the difficulties he, and so many others in urban farming on this scale, have faced. His honesty about the problems and the mistakes they originally made is both illuminating and reassuring. I find it fascinating, after reading *Street Farm*, to read about the amazing progress and durability of Michael's urban farm in *Farm the City*, and how many people are being helped. Experiential capital is indeed vital, and he has it in spades, buckets, and boxes! There are many out there growing food in the city, but few have Michael's experience and ability to understand problems that keep arising. I salute him. This book will help so many urban farmers. I urge you to read it.

— Charles Dowding, gardener, writer, teacher

Whenever Michael Ableman sees a barrier, he runs over and kicks it in. Lucky for us, this strikingly focused anarchist writes about it too, sharing the deeply moving story of reclaiming land and building real community in the most unlikely places, from the ground up.

— Dan Barber, chef/co-owner, Blue Hill and
Blue Hill at Stone Barns, author, *The Third Plate*

The goal of this five-acre network of four farms—begun in the poorest postal code in Canada—is to produce, from thousands of boxes of planted dirt, not just delicious food but salvaged lives. Candid about the difficulties of creating flourishing farms on hot pavements and of making reliable farm workers of dispirited locals who struggle not only with poverty but with assorted personal demons, Ableman has written an important, inspiring, and bravely honest book.

— Joan Gussow, author, *Growing, Older* and *This Organic Life*

Sole Food Street Farms is living proof that creative social enterprises, thoughtful land use, and green jobs can combine to make cities more inclusive and resilient. Michael Ableman's work and passion helped make Vancouver a global leader in urban food systems, with happier and healthier people.

— Gregor Robertson, former mayor, Vancouver, British Columbia

FARM *the* CITY

FARM
the CITY

A TOOLKIT *for* SETTING UP
A SUCCESSFUL URBAN FARM

MICHAEL ABLEMAN

new society
PUBLISHERS

Book design by Meg Reid.
All images © Michael Ableman.

Printed in Canada. First printing April 2020.

Inquiries regarding requests to reprint all or part of *Farm the City*
should be addressed to New Society Publishers at the address below.
To order directly from the publishers, please call toll-free (North America)
1-800-567-6772, or order online at www.newsociety.com

Any other inquiries can be directed by mail to:

New Society Publishers
P.O. Box 189, Gabriola Island, BC V0R 1X0, Canada
(250) 247-9737

Library and Archives Canada Cataloguing in Publication

Title: Farm the city : a toolkit for setting up a successful urban farm /
Michael Ableman.

Names: Ableman, Michael, author.

Description: Includes index.

Identifiers: Canadiana (print) 20190231920 | Canadiana (ebook) 20190231939
| ISBN 9780865719392 (softcover) | ISBN 9781550927320 (PDF) |
ISBN 9781771423281 (EPUB)

Subjects: LCSH: Urban agriculture—Handbooks, manuals, etc. |
LCSH: Sustainable agriculture—Handbooks, manuals, etc. |
LCSH: Sole Food Street Farms.

Classification: LCC S494.5.U72 A25 2020 | DDC 630.9173/2—dc23

Funded by the Government of Canada | Financé par le gouvernement du Canada | Canadä

New Society Publishers' mission is to publish books that contribute in
fundamental ways to building an ecologically sustainable and just society,
and to do so with the least possible impact on the environment,
in a manner that models this vision.

CONTENTS

CONTENTS

ACKNOWLEDGMENTS

This book, and our work, is made possible by the support of many groups and individuals.

The Real Estate Foundation of British Columbia generously funded the writing and production of this book.

Our former operations manager, Lissa Goldstein, weighed in early on the text, and Josh Volk provided important detail for the crop planning section. Carey Jones did incredible work editing, helping with structure, organizing, and providing key guidance in completing the project. All of their feedback and suggestions were invaluable.

We'd like to thank the following individuals, organizations, and foundations who make our ongoing work at Sole Food a reality: Berman Family Foundation; British Columbia Automobile Association; Central City Foundation; City of Vancouver; Concord Pacific Foundation; Diane & Daniel Vapneck Family Fund; Face the World Foundation; Gaia Green Foundation; Giftfunds Canada; Rory Holland; Sadhu Johnston; Timothy Kendrick; Knifewear Inc.; Laucks Family Foundation; MODUS Planning, Design & Engagement; Nickle Family Foundation; Rob and Ruth Peters; The Radcliffe Foundation; Salt Spring Coffee; Social Venture Partners; SpencerCreo Foundation; Swift Foundation; Tides Canada; Vancity Community Foundation; Vancouver Foundation; Vidalin Family Foundation; Wettstein Family Foundation.

INTRODUCTION

S ole Food Street Farms was started to address two challenges—a social one and an agricultural one. Could the simple
act of providing meaningful work through growing food
in the city help folks dealing with long-term addiction, mental
illness, and material poverty? Was it possible to create viable and
credible agricultural enterprises on pavement or contaminated
land in the heart of our cities?

There are many excellent examples throughout the world of
garden-scale and personal food production in our cities—in
front yards and backyards, on rooftops, in alleys, along railroad
tracks and boulevards, and in community gardens. Humans are
incredibly resourceful when it comes to basic survival; we also
have a fundamental human drive to plant seeds, nurture soil,
grow plants, and share the bounty. But in spite of the fact that
the words "urban" and "agriculture" are now commonly used

together, there are few examples that are production scale and truly agricultural.

For over a decade, Sole Food Street Farms has attempted to demonstrate that half-acre and larger plots of paved and unpaved urban land could provide production quantities of food, create full-time jobs, feed neighborhoods, and become successful economic enterprises. We farm on more than four acres of pavement, using thousands of growing boxes, and have a large urban orchard that produces persimmons, figs, quince, apples, pears, plums, and cherries. We harvest an average of 25 tons of food annually. We employ 20 people, have paid out several million dollars in wages, and, according to two university studies, for every dollar paid to our staff there is between $2.25 and $5.07 in savings to the broader society in the form of a social return on investment (SROI) (see page 92 for more on this).

Some parts of our grand experiment have worked; others have not. In that sense we see ourselves as another link in a 7,000-year agricultural history of trial and error. The difference is the urban context in which we work and the social mandate we have. We know that the majority of urban land in the world is either paved or too contaminated to safely grow food, and we have attempted to create strategies to operate within those challenges and to do it on a scale that is significant. We have worked with city governments, landowners, and the broader community to accomplish something truly remarkable that, to our knowledge, has never been done on this scale before.

Now we want to share what we have created with other urban farmers, city planners, and those who work in social services and with underserved communities. The pages that follow make up a "primer" or "tool kit" of sorts. We will discuss our techniques and philosophy and offer information on fundraising and marketing strategies. We will provide helpful documents like budgets, lease proposals, and planting and harvest plans. Finally, photographs of the farms we have created and the individuals with whom we work will help bring to life the realities, challenges, and rewards of this type of endeavor. We hope that our successes and, more importantly, our failures can provide a stable foundation for your efforts.

Many of the resources discussed in the pages that follow are available on our website; please visit solefoodfarms.com /farming-the-city/resources. To view or download, use the password farmingthecity_2018.

For a deeper look at Sole Food's innovative agricultural work and its social mission, including many stories about the challenges and triumphs of our staff members, check out the book *Street Farm: Growing Food, Jobs, and Hope on the Urban Frontier* (Chelsea Green Publishing, 2016).

CHAPTER 1: PLANNING

MISSION

Establishing a simple mission statement is essential for clarifying to both yourself and the world who you are and where you are going. A mission statement defines your goals and provides a reference point for everything you do as a business or organization. We have two mission statements—one for Sole Food Street Farms, which is the farming entity, and one for Cultivate Canada, which is the charity that wholly owns Sole Food Street Farms and oversees its operations and activities.

Sole Food's mission is to empower individuals with limited resources by providing jobs, agricultural training, and inclusion in a supportive community of farmers and food lovers.

Cultivate Canada is a registered charity established to demonstrate and interpret the vital connections between farming, land stewardship, and community well-being; to model the economic and social possibilities for small- and medium-scale urban and rural agricultural and forestry projects, to address disparities in access to healthy food and the knowledge to produce it, and to nurture the human spirit through public programs, classes, and events.

●An aerial view of Sole Food's headquarters farm

EXPERIENTIAL CAPITAL

There is a general belief that access to land is the biggest challenge for new farmers. I'm not sure I agree. While it is true that land—especially urban land—of any significant scale can be challenging to access, I think a greater challenge for beginning farmers is access to the knowledge and experience required to farm well and to run a farm business—otherwise known as experiential capital.

Experience can only come with time. Devoting several years apprenticing or working with an accomplished farmer is an excellent investment for any wannabe farmer. There is a long but somewhat vanishing tradition of this in other parts of the world, especially in Europe and Asia, and in other trades, like building and cooking. Although farm apprenticeships are available in North America, they are often not given the respect they deserve, and are fraught with unrealistic expectations, both on the part of apprentices and the farmers who are mentoring them.

That said, an extended apprenticeship period is the only way to gain the necessary experience to start your own farm and not have it crash and burn within the first few years. An extended apprenticeship period also allows you to work through numerous romantic ideas about farming, come to terms with the day-to-day realities of the work, and find out if it is really a profession you want to pursue. There have been many times when someone comes up to me and acknowledges that I have inspired them to buy a farm or start farming, and I am never quite sure whether to offer my congratulations or my condolences. Taking a few years to learn on someone else's farm allows you to make mistakes that a new farm could not recover from, learn what crops you like to grow, clarify your goals, and develop the skills required to succeed.

When seeking an apprenticeship, look for someone who has been farming for far more than 10 years, whose products are well respected; whose style, philosophy, scale, and approach aligns with where you want to go; and, last but not least, who is willing to have you learn at his or her expense. This will likely not be an urban farmer, as there are so few who are operating on a production scale or who have significant experience. When we started Sole Food, we relied on my experience in both large and small-scale rural and urban production, and applied it to the city. Seek out a rural or peri-urban grower (one who farms on the immediate outskirts of a city) who is well established and invest the time in learning from him or her. Devote several years entirely to learning and to gaining experience, putting skills and experience above financial gain. This may be the single most important investment you make.

FINANCIAL CAPITAL

Another hurdle for a beginning farmer is accessing the capital to purchase tools, equipment, irrigation supplies, seeds, etc., to carry a farm operation through the first three to five years that it takes to become stable. The startup costs can be significant depending on the size and scope of your operation—view a rough budget we drew up at Sole Food's beginning at sole foodfarms.com/farming-the-city/resources (password Farming thecity_2018). While traditional sources of funding such as banks and similar institutions abound, it can be difficult to obtain a loan for a farm operation from these sources as the rates of return and financial profile of farms often do not fit into their normal criteria and expectations. Individual lenders and sources such as crowdfunding may be more receptive to the unique nature of small farm enterprises, especially if they have a social goal.

Because of our broader social mission, our employment model, and the fact that we are wholly owned by a registered charity, Sole Food is able to access funds through foundations and individual donors. If our only goal were an agricultural one, we would hire staff with those skills, as most farms do. Although we endeavor to operate like every farm, supporting our budget through the sales of our products, our social mandate and the challenges of the people we employ will never allow us to operate on the same playing field as other farms. Currently, about half of our annual budget is supported by the sale of farm products, and the other half through fundraising. Though the social benefits we accrue do not appear on the bottom line of our financial reports, they are significant.

Seeking a loan from a bank or financial institution requires that you prepare a business case that demonstrates that your operation will be financially viable and that you'll be able to repay that loan. I often think that writing those documents is like a test, a hurdle that lenders want you to go through to demonstrate that just by preparing such a document you have demonstrated the perseverance and tenacity that will also make your business work. The details presented in the document may be secondary to the fact that you actually created it.

A few good resources for farm business planning are:

Beginning Farmers: http://www.beginningfarmers.org/farm-business-planning/
Cornell University's Small Farms Program: http://smallfarms.cornell.edu /plan-your-farm/planning-funding-your-farm-business/sample-business-plans/
The Spruce: https://www.thespruce.com/write-a-small-farm-business-plan-3016944

Private loans from individuals, family members, or friends can be easier to negotiate and obtain, but can also be fraught with personal and relational challenges if those loans are not backed up with well-written agreements and repaid like any other loan.

It is essential when preparing any business case based on farming to be extremely conservative in the expectations you present. Unless you are a veteran farmer (and even if you are), there will be a steep learning curve, and you can expect it will be three to five years before your enterprise begins to show a return. Income projections should also reflect the vagaries of an unpredictable climate, of changing markets and prices, and of biological conditions that are both within and outside of your control.

Because of the social mandate of Sole Food and its charitable goals and umbrella, we are able to solicit and receive tax-deductible contributions. However, a nonprofit or charity structure requires a heightened level of diligence and accountability when it comes to honoring the support from donors who stand by the organization's mission, and in staying true to that mission. Soliciting charitable funds is an art and a responsibility.

Every year we produce a simple report which presents some of our key accomplishments, some quotes from our staff, a snapshot of our finances, and some thoughts and projections for where we are going. View some of our past annual reports at solefoodfarms.com/farming-the-city/resources (use the password Farmingthecity_2018).

FUNDRAISING PRINCIPLES

- Raising money is about personal relationships: People give to other people more than to ideas or projects (even great ones).
- Think of fundraising as providing an opportunity for a donor, rather than a solicitation.
- Successful fundraising is best accomplished by the people who are most involved with and passionate about the work at hand. Professional fundraisers may be able to design campaigns or strategies, but they can never achieve what those in the trenches can.
- Familiarize yourself with the history of a particular donor or foundation. What do they like to support and in what amounts? Once you are clear about what capacity a particular donor has to give, ask at or slightly beyond that level.
- Fundraising events are often not the best way to raise money. They are wonderful for making friends, celebrating organizational achievements, and educating the public. Too often, though, they cost more to organize and put on than they

actually generate. Host events, but be realistic about what they can achieve.

- After you have received contributions, thank donors numerous times—and in various ways—and stay in touch with them.

SELECTING AND ACCESSING LAND

Finding and accessing land to set up your operations in a city can be a slow and, at times, difficult process. Begin your land search as soon as you have a solid business plan written and are seeking funding. Here are some thoughts on the process.

Farmers look at potential land with an eye to existing soil fertility, drainage, weed pressure, exposure, wind, access to water, access to supplies, proximity to markets, and so on. We train ourselves to read the land, understand what it means when certain weed communities are predominant, observe the growth habits of existing plant material, and employ both sensory observations (smell, feel) and laboratory analysis of native soils.

These instincts are valuable, but when considering a farm in the city, there are often the added constraints of contaminated soils, pavement, and extreme space restrictions. These constraints require an entirely different approach to more traditional food production strategies and techniques, and necessitate that we seek out different elements.

Ideally we look for a site that is a minimum of a half acre (about half the size of an American football field) and relatively flat, is fenced for security from theft and vandalism, is not shaded by high-rise office or apartment buildings, can be easily serviced by trucks or forklifts, and, assuming that the native soil is not usable, has land that is "capped" (covered) or paved over in order to avoid the health, legal, and permit issues that relate to soil contamination. (Many municipalities now provide soil contamination testing services for new farm or garden endeavors.)

● *The original wood growing boxes before filling them with soil at our farm at Pacific and Carrall Streets*

● *Our urban orchard at the corner of Main and Terminal Streets*

Urban land typically has a much higher value (in the narrow economic definition of the word) than the income that could be gained from farming. And so we seek out land in industrial parts of the city; land in areas that are less likely to be developed; land that will be tied up in legal, permit, or development issues (and therefore go unused) for a minimum of three years; or parking lots that are not fully utilized.

I have always believed that ownership is highly overrated as the dominant model for young and beginning farmers. Long-term leases (a minimum of 10 years) for open-field, rural farms are a better way to get started. It normally takes five years to get to know a piece of land, improve the soils, understand how the light and air moves across the land, and establish stability in a new farming operation. However, for commercial urban farms, those time frames are unrealistic; urban land is just too valuable for any owner to tie up in anything more than a shorter-term lease. Situations that have the potential to provide a longer tenure are ideal, but we settle for less.

Finding a site that has sympathetic ownership is also important. A farm on urban land is not a typical use, and requires a landowner who is open-minded and willing to consider new ideas.

One of the most important aspects of selecting an urban farm location is learning about a potential site's neighborhood. Who will your farm neighbors be? What ethnicity and cultures are represented, and what do they like to eat? Will a farm in the neighborhood be well received, or will there be resistance? How will neighbors respond to people coming to work each day, occasional noise from machinery, a possible sales outlet, composting activities, and different smells and sounds?

Often urban land selection decisions are dominated by the simple fact that land in the city is limited in availability. This is especially true for cities along the coasts where the price of land tends to be quite high. There is a very different conversation going on about food production in midwestern U.S. cities, like Detroit or Milwaukee, where economies are struggling and property values are low. Food production models within those environments will be different than those in the higher-valued real estate markets of places like Vancouver, where we farm. Just like rural farm fields that each have unique personality traits and require different approaches, urban farm spaces and the neighborhoods they inhabit demand that we consider a host of factors before beginning in order to be successful.

When seeking out a potential site to lease, it is helpful to have prepared a simple proposal that outlines who you are, how you intend to use the land, and what benefits you will provide to the landowner. This document provides an important introduction to landowners and demonstrates your intent and that you will be a solid and dependable renter. Visit solefoodfarms.com /farming-the-city/resources for a look at a lease proposal we've used to present to potential landlords (use the password Farming thecity_2018).

Whether you are farming on municipal or private land, the landowner—a city, a company, or a private citizen—will typically be responsible for drawing up the lease. While leases are normally pretty straightforward "boilerplate" documents, there will often be language added in to address the specifics of your arrangement. If possible, it is always best to have a lawyer review any lease documents before you sign them.

PERMITS AND POLICY CHALLENGES

In spite of all the attention the concept is getting, urban farming is still not an activity fully accepted by and integrated into municipal zoning. Most cities' regulations were created to accommodate factories, roads, businesses, and homes or apartments—not to grow food.

This can also be an issue when cities expand into what were once rural areas. I experienced this acutely in the 1980s when the 12-acre California farm I was running became totally surrounded by urban and suburban development. Complaints started flooding in to local authorities about common farming activities that we had always done. The crow of roosters, the smell of compost, the sounds of tractors and other farm machinery, and the signs advertising our products were all foreign to the new neighbors who moved in around us. Even though our farm had been in operation for over 100 years, we had to defend our right to exist.*

Since farming is not on the books of most municipalities, the permit process to carry out farming activities can be challenging and, in some cases, prohibitively expensive and time-consuming. This happens even in cities that have publicly declared urban agriculture a goal, Vancouver being just one example.

Conditional use permits, variances, and building permits for simple structures like tunnel houses for season extension, environmental reviews and studies, contamination mitigation and strategies, site plans, development permit applications, community input, operational management plans, and many other processes and bureaucratic demands can consume vast amounts of time and financial resources. And in many cases, they are to fulfill requirements that are disproportionate to the activity that is at hand. For example, a tunnel house is a steel-frame structure covered with a sheet of 6-millimeter plastic; it is used for extending the season and for growing heat-loving or cold-sensitive crops. It is a simple structure, yet the City of Vancouver's building department required that we create fire exits and complex air circulation systems, and hire structural engineers, and more, all to fulfill the one-size-fits-all urban municipal bureaucracy that was never designed to regulate farming.

Another example: The shipping containers we use at our farm sites to secure tools and equipment and provide office, refrigeration, and storage space were required to be anchored into the ground with steel and concrete. When empty, those containers weigh almost three tons and are always the only thing standing after any tornado, hurricane, or flood.

* The book *On Good Land: The Autobiography of an Urban Farm* (Chronicle Books, 1998) and the film *Beyond Organic: The Vision of Fairview Gardens* (Bullfrog Films, 2000, narrated by Meryl Streep) provide more information about this project.

● *Sixteen thousand square feet of high tunnel houses where we grow tomatoes, peppers, eggplant, strawberries, and melons*

These permitting processes are always more cumbersome and difficult for those who are first in line pioneering a new effort. And while the City of Vancouver, for example, has begun to address the challenges of urban farms by creating new policy, it is still geared to very small-scale production units, not the type of larger farming enterprise that we represent. Many cities are now making an effort to write municipal code that will help these endeavors, but we have a long ways to go before these types of land uses are fully accepted and intelligently integrated.

Environmental concerns are also a reality in many urban farm sites, due to contamination from historical land use. For example, one of Sole Food's farm sites is on land previously occupied by two former gas stations, and on top of what used to be an industrial shipping and rail yard (see Soil/Fertility, page 20, for more information on testing soil for contamination). I suspect that anyone who attempts to undertake urban farming on any significant scale will only be able to access or afford land that has some form of abuse or sordid history. We take seriously the need to grow our food responsibly and ensure that our products and our staff are safe. We predominantly farm on pavement, meaning we must use soil-filled containers to grow our food, which completely isolates the growing medium from any contamination and allows us to farm on impervious surfaces like pavement. Still, we have had to jump through some major hoops both to grow safely and to get the permits necessary to do so. For example, the City of Vancouver required we get an environmental assessment done on our sites before we began farming. We partnered with a Canadian environmental consultancy to produce a 58-page report that included a methodology by which we could safely farm on brownfield sites, and a rubric for assessing the viability of future potential sites.

CHAPTER 2: GROWING

INFRASTRUCTURE

High land values and contaminated or paved-over soils are core challenges that exist in most urban areas. And while there are situations that allow for growing in native soils, they are the exception.

In-ground growing is far less expensive—no costly containers and imported soil are required—but the responsibility associated with insuring that native soils and the food that comes from them are not contaminated is huge. Clean urban soils are rare, and even relatively clean areas may have contamination pockets that are difficult to detect. Even in those situations where native soil is clean, the investment in improving the fertility of that soil may not make financial sense unless the lease is long-term.

At Sole Food, we have designed a container growing system that isolates the growing medium from contaminated soils, allows for farming on pavement, and is movable on short notice (due to the common short-term tenure on urban sites). While our container system is constantly evolving, the basic concept addresses the three core challenges of urban agriculture in one elegant solution.

● *Filling the original wood boxes with soil at our farm below a sports stadium*

In pursuing the perfect container for growing food, we went through an amazing number of ideas, possibilities, and design and construction incarnations. We wanted a box that was large enough for efficient production, movable by forklift or pallet jack when full, deep enough to provide a substantial rooting zone yet not so deep as to drive soil and material handling costs up, resilient to the elements, stackable and nestable, impenetrable to rodents, and inexpensive.

As is often the case, cost became a primary constraint during the early phase of our project, and so we initially looked at containers that could be repurposed. We considered the plastic bulb boxes from Holland that farmers use for storing and transporting produce. While there are millions of them out in the world, we determined that they were too small. Next, we considered collapsible, waxed cardboard watermelon and winter squash bins, but they proved too deep and not durable over time. We had prototype containers built from the heavy fiber material that came off a collapsed roof of a Vancouver sports facility, and we tried commercial fruit harvest bins, large bags, pots, and PVC pipes.

On our first experimental location, we built containers using dimensional lumber, but realized they were too expensive on any significant scale, would eventually decompose, and were difficult to move. Finally, we settled on pallet collars and used shipping pallets for our initial site expansion on our largest production site. A pallet collar is a collapsible, hinged, bottomless plywood box that slides over a shipping pallet to create a movable open-top box. Every community has an abundance of used shipping pallets, stored in shipping yards, warehouses, and onsite at trucking companies. We liked that the collars could be manufactured and shipped in large quantities for a fraction of the cost of a new plastic box.

Unfortunately, we discovered that untreated plywood pallet collars do not hold up to the elements. In fact, ours started delaminating within the first year, and the rats took up residence in the easy-to-access pallets underneath, feeding happily on our vegetables.

This was a large-scale failure. We took a huge financial and production hit, and realized that having boxes fabricated to our specifications was the best route, to use as replacements as well as for new site expansions. We began working with a local plastics manufacturer to design containers to our specifications.

● *Delaminating wood growing boxes–an expensive lesson*

First, we designed plastic boxes for a one-acre urban orchard. These containers had more depth to accommodate the deeper rooting requirement of the trees. We designed the boxes to have a lip around all four edges to support hoops for season extension and forklift tabs underneath to allow for complete movability.

The boxes for the trees were virtually indestructible. I tried with a Bobcat tractor—running over it with the crawler's tracks, crushing it with the bucket, and applying every other form of abuse I could come up with. The box survived, did not crack or break, and was put back into service after regaining (with some help) its original shape.

Our next goal was to replace the 3,000 original (now decomposing) pallet collar vegetable boxes on our largest site with plastic boxes. The original "premium" vegetable growing box we designed incorporated every bell and whistle—it was double-walled to provide insulation from excessive heat, had a reservoir for sub-irrigating, tabs to allow the boxes to be stackable when full and nestable when empty, forklift slots, a lip around the top with holes for supporting hoops, and the ability to interconnect drainage so that spent irrigation water could be collected and recycled.

● *Our latest design for our plastic vegetable growing box*

Some of these features—the double-walled construction and the internal reservoir in particular—were eventually eliminated due to cost. We tooled and fabricated these new boxes en masse to replace deteriorating wooden boxes at three sites. After three years of growing in them, the plastics solution appears to have been a success. The farms are productive, the soil is contained, and we have fewer issues with rodents. In addition, the boxes have created interest in the wider world outside our farm—in fact, we get more requests (from individuals and organizations) for purchasing the boxes than we do for the produce that grows in them! As a result, we are looking into setting up a winter enterprise for our staff (who are either laid off during the winter or have their hours reduced) to manufacture and distribute the boxes.

SOIL/FERTILITY

All of us who possess a personal ecological ethic, and try to be true to that ethic, know that our modern existence is fraught with endless paradoxes. Every time I get out of bed, pour a bath, turn on the lights, or eat a meal, I am participating in a vast ripple of repercussions. I try to make choices that have the lowest impact, but I also accept the duality of my existence as I navigate through life's daily contradictions while holding onto a larger social and environmental vision.

Soil is the foundation of what we do as farmers, and although most of our society treats soil like dirt, every one of us is inextricably tied to it and dependent on it. I have spent most of my career in farming working on more typical open-field farms. In those more traditional situations, we start with existing soil and constantly work to make it better. However, it's not so simple when we are farming on pavement or having to isolate the growing of food from contaminated urban soils.

Our original plan at Sole Food was to make soil by composting Vancouver's waste organic materials, like kitchen scraps, uncontaminated yard waste, and restaurant waste. We designed a system and came up with a budget, but shelved it after realizing how hard it would be to get permits, the difficulty in finding a site that would allow waste recycling, and the limitations on our own time and financial resources. The infrastructure, equipment, and management required to produce roughly five thousand tons of soil for five acres of boxes was too much for a new farming project that already had enough other challenges.

As a result, we had to rely on purchased and donated soils. We recognized that it requires an enormous amount of fossil energy to move the vast amounts of raw materials it takes to make soil, to screen and separate it, compost and blend it, and to truck and distribute the final product. We had to accept the financial and ecological costs of accommodating a farming system that attempts to produce commercial quantities of food on pavement in boxes. We knew that we could not safely grow food in contaminated urban soils or physically do it on pavement, so we had to start by acknowledging the imperfections of the system as we tried to create the right soil foundation for our farms.

If we weren't going to do it ourselves, who had a product that would meet our need for soils that were biologically alive, fertile, and affordable? We were looking for high-quality compost blended with sand or peat to increase drainage. We built a relationship with a nearby composting company to develop a product that could meet the needs of containerized vegetable production. In this process, we reviewed soil analyses and visited their facility. An analysis from any soil company you're working with is a must. If you are working with containers, you will need a soil that breaks down slowly so that you don't lose volume too quickly, has a balanced mix of the macro- and micronutrients that vegetable crops require, and drains well, but not too well. Your soil will be the basis of everything you do. Regardless of what happens with leases or infrastructure, it is your most valuable resource over the long term.

Growing food on a commercial scale in boxes presents a completely different situation than in more traditional open-farm fields, where chemistry and biology and fertility are able to play out in a vast and complex network of relationships. Soil in boxes is cut off from that broader dynamic environment, and biological relationships are limited. While we try to infuse that biology into the boxes, our efforts will always be imperfect.

Starting with fertile soil can give you a false sense of security. If you're not careful, soil energy can be easily depleted. The challenge is in maintaining existing static fertility over the long haul. In our early years growing in containers, the new, rich, loose soil provided wonderful results. Plants and leaves and roots and fruit grew with abandon. Our raised boxes received heat from the sides, were well drained, and provided ideal conditions for plant health.

But after a couple of seasons, we could see things changing. The organic matter content of those boxed contained soils was disappearing, leaving behind a soil structure that was more like powder. In spite of our efforts to inoculate the soil after each crop with worm castings and compost, and add mineral fertilizers, we found the soil biology and fertility was difficult to maintain within the contained and constricted environment of a box.

We discovered the hard way that many standard practices used in open-field agriculture, such as cover cropping, were far more difficult and in some cases impossible to implement in container-based systems. There were basic physical challenges in incorporating crop residues and green manure cover crops into containers. We tried, but unfortunately, in our containerized system, without a mechanized alternative for tillage, this work all needed to be done by hand. On a scale of more than an acre, hand work becomes daunting and difficult to sustain.

● *Harvesting radishes*

All good organic practice includes well-planned crop rotation to prevent pest and disease pressure and to maintain soil fertility and manage weed pressure. We were challenged in implementing good crop rotations as well because we simply did not have the time and space. In addition, we discovered that limited soil depth and the dramatic soil temperature fluctuations that were occurring in our raised containers on pavement were problematic for some crops. Containers also limited our ability to use machinery for tasks such as tillage and mowing, and when we were able to till, it seemed to have a more dramatic negative effect on soil structure.

On the positive side, containers have virtually no soil compaction, excellent drainage, earlier production, and more rapid fruition (due to more heat reaching the raised root zone). Boxes also have physical and psychological advantages for farmers; crops are higher off the ground and so require less bending over.

Finding the optimal dimensions and soil depth for container production is critical for ease of working, root depth, plant health, and nutrient reserves. The larger the box, the greater the soil reservoir and uninterrupted working space, but the more expensive it becomes to manufacture, fill with soil, and move if necessary. Balancing the financial challenges of an expensive container system with creating optimal conditions for working and for soil and plant health is an ongoing challenge.

As a starting point for knowing your soil, apply basic observation skills: What does the soil feel like? How does it smell, or even how does it taste? How does it react to dry or wet conditions? How are the plants performing? Simple observations can provide a critical feedback loop and ground your soil experience.

Submitting soil samples to a lab for analysis provides more precise data that will help round out your observation and allow you to design a more detailed soil management program. (Every region has established laboratories that will test your soils and provide a complete written analysis; check with your local agricultural organizations or extension office for the names of these labs.) For the most accurate test, we take small samples from different areas of the farm and mix them together. We want a clean sample, without any large chunks or bits of organic matter. We do this for each of our sites and submit it to the same lab every season. The soil test won't provide the perfect answer to a fertility puzzle, particularly in a containerized setting where the soil can be slightly different in each box or between different sections of boxes, but it provides a guide for each site and a baseline from which to work. Like any experiment, you must have a control to analyze the data accurately. In this case, we test at the same time every year, and we use the same lab for analysis.

Testing at the same time each year is vital because the fertility within the soil and the soil chemistry change with the seasons.

What we look for with the soil test is how things are changing. The soil test will give you an absolute number, but we look at it as a snapshot and range of where we are at that moment in time. Are we generally low on calcium? Are we okay on calcium? Is the soil very high in calcium? The exact number is going to vary across the seasons because rainfall, biological activity, plant growth, and warmth can all impact test results. There's no one good time to do the test; what is most important is performing the test at the same time of year each year, so we have an honest comparison. From this soil test, we use our observations and our intuition coupled with the recommendations of the lab to create a plan for maintaining long-term fertility and soil health.

Within all of the constraints of containerized crop production at Sole Food, we have been evolving a system that combines the use of concentrated mineral and powdered organic fertilizer mixes and compost extracts or teas, with the addition of purchased compost.

While the urban environment presents an unlimited supply of raw materials for making compost, the regulations, health and odor concerns, difficulty in finding a suitable composting location, and the handling challenges of large volumes of organic materials make it a nearly impossible endeavor on a significant scale. Enclosed, self-contained composting systems are a great option, but the capital cost of those manufactured units is high and, as such, prohibitive for a start-up farm project.

The use of compost tea is based on the recognition that soil fertility is about much more than chemistry, that soil biology is the engine that drives the system. And although we recognize the physical conditioning benefits of compost applications, using compost tea extracts provides a far more efficient alternative to the constraints of hauling and applying large volumes of physical material within an urban environment. Extracting, transporting, and applying microorganisms require far fewer resources and energizes the soil. Ratio feed injectors are wonderful for this purpose; they inject minute amounts of compost tea into irrigation lines with every watering and are driven purely by water pressure.

Foliar feeding with a kelp solution is also an important part of our fertility plan. This type of feeding provides a micronutrient boost through the stomata on the leaves of the plants. Foliar feeding is generally best done in the evening or with sufficient cloud cover or cool weather; if you apply the solution during the heat of the day, you may not get the same results. Results can also vary by plant—some plants are better at absorbing through their leaves than others.

At Sole Food, we rely on minerals and other powdered amendments in a dry concentrated fertilizer mix that we apply before each planting. This mix may include glacial rock dust, greensand, rock phosphate, alfalfa meal, bonemeal, blood meal, feather meal, mined potassium sulphate, lignite, kelp meal, and gypsum.

Every agricultural system has ongoing ecological compromises; our urban agricultural system is full of them. All so-called natural fertilizers are either mined or harvested from some distant environment at varying levels of ecological and financial cost, and, in some cases, social expense. While every farm, rural or urban, faces the reality of having to amend with mineral fertilizers, our ultimate goal is always to view the farm as a self-sustaining, self-feeding living organism. Creating a fertility cycle from within in order to support the health of the farm organism is far more achievable on a rural farm that can support a mix of animal and plant life and the kinds of longer-term rotations that are necessary. Farming in boxes on parking lots limits that possibility.

We continue to experiment with different ideas and approaches to address these fundamental challenges, always believing that we will eventually evolve a more efficient, more ecologically sensible, and more economical approach to maintaining good soil fertility in large-scale box agriculture. But it will always be an imperfect system, established to feed and employ folks who live in an equally imperfect urban construct.

WHAT TO GROW

The answer to the question of what to grow requires some understanding of the community you are serving, existing markets, local climate, scale, availability of skilled labor, access to and cost of water, and your skill level as a farmer.

I always advise to begin by producing the foods you like to eat; you'll naturally do a better job growing them. If you are just beginning, you may want to focus on cutting greens and roots, and expand your product mix as you build your skills and confidence.

Chances are your product mix will be dictated by economics as well as by space. You'll need to assess your desired crop mix relative to its cost of production and its value in the marketplace. The cost of production in the city (due to higher infrastructure and labor costs) will likely be higher. What products can you grow in relatively small spaces that will provide the highest yield and the highest financial return per square foot?

It's important to consider that you can have a real edge when you are the earliest and/or the latest to come to market with certain crops. Being the earliest at the market forges loyalty, as does being the last one to have something. Both early and late often bring higher prices.

It is also essential that you choose a small body of what I call "signature crops" and that you make a commitment to having those products in consistent supply and quality for the full range of your growing season. It is all too common to see a less experienced grower show up with a crop for a week or two, never to be seen again. Your customers want to be able to depend on you, and you'll gain their loyalty when you can show up week after week with a dependable base body of certain products.

Do an informal survey of local farmers' markets, stores, restaurants, and CSAs (Community Supported Agriculture) to find out what is already available in your area. While it is true that you will likely find many similar products, the quality and varieties of those products will vary a great deal. Take that survey information, blend it with your comfort and skills, and add to that information the production space you have available. Next, come up with six to eight items that you feel you can grow consistently over the whole season, given what you know about what others sell. Do not be intimidated by producing something that is already available if you think you can do it differently, more consistently, or earlier or later than other farms.

I also make a practice of visualizing my display tables at the farmers' market or my availability lists to restaurants when I am ordering seeds or producing a crop list. What is the relationship of the crops I am choosing to each other visually, in their culinary use, and in terms of rotation, soil fertility, or space demands? Imagine your display and how it will look while making this list and while ordering seeds. How can I stand out at the farmers' market or in the eyes of a chef? What can I do to differentiate myself from what is already out there?

In our first years at Sole Food, we focused our crop mix on simple things such as radishes, salad greens, chard, kale, spinach, etc. I wanted to ensure success for our staff, and so we kept our crop mix simple and limited to those items that were reliable and bulletproof. As our crew gained experience and confidence, we expanded our production diversity, evolving a list of crops that fit the demands of the chefs we work with, the farmers' markets we participate in, and the scale we operate on. Because of the boxes and relatively limited space available to us, we do not try to grow large amounts of super-hungry and space-loving crops such as corn, winter squash, etc.

Finally, always include a few new crops that are purely experimental; this is how you grow for the future. In addition, try to produce a few items that are just for yourself, your family, your co-workers, and your neighbors and friends. Sweet corn is a good example of this. It's a crop that friends and family always love, but a difficult one to rationalize in terms of its use of soil nutrients and space. We also grow fresh lima beans—it's another crop that makes no sense economically, but keeps both us and our friends and neighbors happy.

TIME, SPACE, AND THE IMPORTANCE OF A FARM PLAN

Time and space are two elements that are critical in any farming enterprise, and the smaller the enterprise, the more critical they become.

Within small urban agricultural enterprises, the careful timing of all of our actions allows for maximum production, greater efficiency, higher quality, and more success in the marketplace. Time in the small farm enterprise must be balanced with space. They are interconnected.

The successful urban farmer will develop an overall plan that plots out the stages of a crop's development from seed to

● *Crop production and diversity at our largest farm*

sale, including planting information, harvest details, and sales/
marketing projections. The plan determines, for example, how
many carrots will be planted, in which block and during what
time period, and when and in what amount repeat plantings will
happen. And while instinct, experience, and intuition are critical
to farming, it is helpful to have a good written plan to guide you,
especially if you are just getting started.

Farmers typically fall into one of two distinct personality types:
those who rely more on intuition, experience, and memory and
those who rely more on planning. Find a way to develop and
incorporate both aspects. Making a good farm plan each year
is about building the foundation. Intuition and experience will
allow you to respond to the inevitable changes of a biological
system that is constantly shifting. The plan is the guide, and as

the season progresses, a good farmer knows how to move within that plan and when to deviate from it.

Before making your plan, revisit your mission and goals—whom are you serving, and what is the purpose and vision of your farm? Let these goals guide you as you develop the plan, and again as you implement it.

If you are growing for a membership program like a CSA, a good plan is very helpful in order to ensure a consistent supply of a variety of foods to that membership. This also holds true if your crop system is very diversified: a large number of different crops, plantings, and harvests over the course of a season are more easily managed with a strong plan.

The planning process also serves to even out the workload over the course of the season. Some people may have a goal of working in very distinct chunks: for example, just hiring in a crew for a few weeks and having the farm be quiet for the rest of the year. Others may want to keep the work evenly spread out over the course of the season. If you are working by yourself, spreading out the workload helps prevent burnout, and if you work with a crew, providing them with consistent predictable work throughout the full season helps retain people so that you don't have to retrain folks every year.

Good planning helps you use time efficiently during the season and anticipate needs well in advance in order to prepare for them; good planning prioritizes proactive management over reactive management. For example, reactive management could be planting a crop and realizing too late that it needs to be trellised but you don't have the stakes or the string, and that your time is already committed to some other crop. With proactive management, you have thought through all of the needs and cycles of each crop and have recognized how those needs combine with all the other crops. While there will always be a certain amount of stress associated with commercial farming, good planning can reduce that stress to a manageable level.

CREATING AN ANNUAL FARM PLAN

Your farm plan will likely end up being a series of sub-plans: documents and spreadsheets that, used in tandem, help you track your crops from seed to sale. The overarching plan should include at least the following elements:

• What you are going to plant
• How much of it to plant
• Where you will plant it
• How you will plant it
• When and how you will harvest it
• How much it's going to cost to grow it
• How much the crop is going to yield
• How and to whom you will sell it
• How much it's going to gross financially

To get started creating the plan, you'll need to have some data on hand, like field notes, yield data from previous years, the dimensions of your planting blocks or fields, and how many boxes or beds can fit into a block or a field. You need to know how much space you have to work with, and any characteristics of those spaces that make them more or less suitable for different crops or for planting at certain times of the year. Other helpful resources are seed catalogs, any notes from past markets or marketing, and any field diaries from prior years. Seed catalogs are an excellent free source of information if you are just starting out and don't have any field diaries or prior notes to fall back on (and if you have been farming for a while, they are a great addition to your own notes). Some seed catalogs, like Johnny's, have great cultural information sections that include yield data, growing requirements, and field spacing for a multitude of crops.

You can create your plan on paper, but a computer has many benefits: spreadsheet programs like Microsoft Excel allow you to insert formulas, cross-reference and calculate costs and projections, and visualize time and space.

At Sole Food, we typically work backward, creating a sales plan and a harvest plan, and then a planting plan.

To create your sales plan (what you will sell, and to whom) and harvest plan (when you will be harvesting each crop for your customers), you'll need to determine what your six to ten anchor or signature products will be, what outlet(s) these products are best suited for (restaurants, farmers' markets, retail, etc.), and when you can have those products available. Start the sales plan with some simple research (visiting markets, talking to other growers, etc.) to come up with a list of products and a price range for each product (keeping in mind that prices will differ depending on your sales outlets). If farmers' markets are your primary outlet, your sales plan must take into account the particulars of each market, including what products customers will buy and in what quantities. Where do the people who shop at that market come from? What do they like to eat? If you want to sell to restaurants, your plan will be based on their particular menus and favored ingredients, and the philosophy, style, and tastes of the chef.

Once you've decided on your products and sales outlets, estimate how much you think each outlet will take of each product. A good rule of thumb is to grow as much as you think you can sell plus a little more, both in quantity and variety. The "little more" is important, as it represents that constant optimism that is required to marry the biological nature of growing food with a market economy.

Your harvest plan will start with the end product (for example, radishes) and work backward to make some clear harvest projections. The goal is to create a crop versus time chart and design the harvest based on what you'd like to be selling and when. Ask yourself, How much space do I have? How much labor do I have? Each of these elements has an associated time component. For example, you might have labor that's available at a certain time of year but not at another time of year. Or you might have space available in the spring that won't be available later in the summer. Managing these different factors will help you create a realistic harvest plan (and help you ensure that you'll be able to grow the products in the quantities you've planned).

Using the sales plan and harvest plan, you can create weekly yield projections for each of your primary crops, allotting percentages of those yields to each type of sales outlet.

To create your planting plan (a large spreadsheet that lists all your crops and their planting details), you will need to input some basic data including yield and space available. Having a standard bed length and width keeps a planting plan simple and allows for easier rotation of crops. Using your own yield data, or that of another farm, you can estimate how much space you need to produce your desired harvest. You can then assign each crop a space in your field and project harvest dates.

Microsoft Excel formulas can help determine how much space to allot for planting and when (on what date) to plant. Using your yield per plant and your preferred spacing, you can calculate how many bed feet you need to plant (and number of rows per bed, inline spacing, etc.). To determine planting date, use your harvest date and work backward based on days to maturity for the given crop.

You should also note whether the crop is direct seeded or transplanted. For transplants, you will need cells for greenhouse information, including how many weeks the crop takes to mature in the greenhouse, what date it needs to be seeded in the greenhouse, how many plug trays you will need to seed it, and what size trays. If the crop is direct seeded, you will want to add a column for the seeder settings, particularly if there is more than one person doing the direct seeding. Which seeder is used for this crop? What is the hole size or roller size? What depth is the shoe set at?

In addition to being integral to the planting process, your planting plan can help manage your seed inventory. At the end of every season at Sole Food, we take a seed inventory that tells us how much we have of which seed and where it came from. Using this inventory and the following year's planting plan, we calculate our seed needs for the coming season. Formulas in your planting plan can calculate seed need based on spacing and planting area. Add product numbers for ease of ordering.

After your harvest plan and planting plan are completed, it's helpful to combine them to create a farm map that includes your field blocks, your crops, and planting/harvest information. Essentially, you're planting the entire farm on paper—consider it a dry run for the real thing. It also ensures that you allow enough space to grow the crops in the volumes you need. It is far easier to go back and reconfigure a plan on paper than to do it in the field!

For a Sole Food farm plan, visit solefoodfarms.com/farming -the-city/resources and enter the password Farmingthecity_2018.

IMPLEMENTING THE PLANS

Getting all your planning down on paper—thinking through and visualizing planting your farm—helps to avoid surprises, project more accurate yields, accommodate better crop rotations, and share all this information with those who are doing the actual implementation. Making the farm map visible and accessible to other members of your staff also allows them to visualize the big picture and imagine the farm over an entire season.

The document also becomes a record-keeping device. You can make a blank map behind the farm map, and use it to record what actually happened in that field. It often happens that what ends up in the field deviates from what was on the plan, and having those maps (both the projected map and the actual map) helps to keep track of what the changes were, and better plan rotations within a season and from season to season.

● *Alain updating seeding charts*

The plans you've created can grow large and unwieldy, so it's helpful to pull out information from them to create a number of smaller documents: for example, seeds that need to be ordered, tillage schedule, a greenhouse seeding chart, a planting chart. These can be sorted by date so they can function as to-do lists; some can be hung on a clipboard so that someone on the field crew can take notes on what was actually planted and harvested (and when and from where).

Visit solefoodfarms.com/farming-the-city/resources (use the password Farmingthecity_2018) for some examples of the documents discussed above.

CROP ROTATION

We rotate crops to support healthier soils; improve quality and yields; reduce pests, diseases, and weed pressure; and cut down on work. Plant families tend to draw on the same nutrient reserves, and moving those families in a predetermined way avoids depleting those reserves and allows soils to achieve more nutrient balancing and cycling.

Crop rotation planning becomes more challenging with smaller production spaces, as the need to return to the same land reduces the time that soil has had to restore. Product diversity also has an impact on rotation plans—the more diverse the product mix, the more challenging and complex the rotation becomes. Large open-field farms have the advantage of being able to use longer-term rotations; in some cases, crop families may not return to the same field for 10 years or more. This is the ideal, but it is rarely possible to achieve on the very small scale of most urban production systems.

You might not always stick with your set crop rotation plan. New pests and disease issues could arise, your crop mix might change, and there will be unanticipated climate and field conditions. Having a plan is important, but being willing to adjust and move away from that plan in order to accommodate the ever-changing elements of biology, climate, markets, and so on is equally important.

Whether you stick with it or not, the plan does help to inform future decisions, makes record-keeping easier, and makes it easier to respond in the moment if something changes—not to mention it aids those of us who cannot possibly keep everything in our heads.

At Sole Food, we start our crop rotation planning by listing out the crops we grow and dividing them into categories. The list includes basic headers, so the crop family on the left, what the crop is, how many beds we had been planting, the season, whether it's on overhead or drip irrigation, etc. We use a spreadsheet to sort by family or number of beds or by season, and we begin to see where there are some dividing lines we can take advantage of.

• *Lyle moving crop residue*

A good place to start dividing is by crop family (brassicas, nightshades, etc.). It is helpful to come up with a standard block size and begin to group your most prominent crops into groups within those blocks. Next, look at what the progression should be, taking into account things like weed pressure, any disease issues that might be in the soil, etc. For example, club root is a disease that affects brassicas, and it requires you keep brassicas as far away from each other as possible in the rotation.

If your block or field dimensions are not similar, you might decide to plant less in a particular year. If you have biennials, you use double blocks. Smaller plantings of multiple crops might go into a space where the previous year you had just one thing that you planted more of.

The goal is to create an optimal progression for weed control, fertility needs, and disease and pest pressure, and to consider edge effects on contiguous blocks from irrigation or neighboring plants shading out each other.

SOIL PREPARATION

On open-field farms, there is a vast array of tractors and implements available to do primary soil tillage, shape beds, and incorporate organic matter. In containers or on very small-scale plots, hand tools are often the only appropriate technology available for the job. At Sole Food, we are working on designing and fabricating an electric device that will surface till an entire growing box at once, but in the meantime, we do that work by hand.

After a crop has been harvested, we remove the organic material from the bed and put it into the compost pile. If we were doing this on a larger farm, using tractors, we would mow, spade, or till the bed and incorporate the organic material where it stands into the soil. But due to our limited space, we must remove that material, compost it, and later return the compost to the box or bed. We also do not have the time or space for crop residues to break down thoroughly before replanting.

Once we remove all prior crop plant residue, we use a digging fork. If the bed is relatively clean and loose, we push the fork into the soil and press back to loosen it, aerating the soil a bit and breaking up any remaining prior crop roots that are still underground. In many cases, especially when turning short-term crops, the bed will still be open and loose and we can skip the fork entirely, apply amendments, and use a rake to loosen the top 6 inches of the soil. If the bed has had a long-season crop in it, the soil may be more compact or have prior crop roots, so we may have to physically turn over the soil with the digging fork. In a "manufactured" soil where there is no real differentiation between topsoil and subsoil, we can get away with mixing the full 16 inches of soil in the box, breaking the soil up and aerating it. This also helps redistribute moisture and nutrients, as boxes tend to be wetter on the bottom and drier on top.

The next step in box/bed preparation is raking. Using a manure rake, we make a second pass over the bed to break up any big clumps and level the planting area. If it's a box or bed that is relatively clean and loose, we may skip this step and rake only after amendments have been applied. Either way, after the bed is raked out and the soil tilth is optimal, we apply amendments, which are distributed evenly on top of the box or bed and then scratched into the top 3 to 6 inches of soil. At this point, the box or bed is ready to be marked for transplants or direct seeded.

• Kenny raking beds in preparation for planting

PROPAGATION

There is no place where time and space play out as profoundly as in the propagation stage of our work. The smaller the scale of the farm, the more critical it is to have a working propagation program that produces high-quality transplants. Rather than waiting for fields or boxes to complete their productive cycle before replanting, plants can be growing in flats while the space where they will be transplanted is finishing its cycle. Not all crops are transplanted—some will be direct seeded—but well-grown transplants provide a significant boost in time, quality, and productivity for any small farm.

There are numerous approaches to propagating plants for transplant. The most common systems are soil blocks and wood and plastic flats, in an array of sizes. We use plastic "speedling"-type flats that are sized for the particularities of each crop, required transplanting timing, germination rates, etc. The cell sizes of flats vary; the smaller the cell size, the faster that cell will fill with roots and be ready to move into the fields. So a 200-cell flat will have a larger rooting area and a 400-cell flat a smaller one. For most of our crops, we use 200-cell flats. This type of flat allows for very fast plant removal and transplanting in the field. We use a high-density flat of 400 cells for a slow-germinating crop like parsley that requires both time and a high number of transplants due to close spacing.

Other crops like tomatoes or peppers will start their lives in small-cell flats, and then be potted into larger cells. This produces larger, more developed transplants and allows cold-sensitive crops to take advantage for a longer period of time of the controlled atmosphere of a propagation facility.

Our well-drained raised growing boxes allow us to plant much earlier than we could in an open-field situation where the ground can stay wet and inaccessible for a longer time in the spring. This means we must start propagation late in the winter, at the end of January or beginning of February. The climate in the propagation house is controlled by small space heaters and by a circulating hot water system that warms the root zones of new seedlings. For example, the ideal soil temperature for germinating nightshades is approximately 70–75°F or 20–25°C, but ambient air temperatures above the immediate propagation growing beds can be quite a bit cooler. Germination tables with built-in hot water circulation are excellent for this. The tables can be filled with moist sand, which helps conduct the heat. If you choose to start your plants very early, when daylight hours are limited, you may need supplemental lighting as well.

In the winter, the artificially warm and moist environment of a propagation house can make plants susceptible to fungal disease. Installing air circulation fans is essential, especially for disease-prone plants like tomatoes. Daily monitoring of plants at the propagation stage is also important, to catch any changes in plant health before they become critical. Careful timing of watering young plants is also vital; the goal should be to keep leaves as dry as possible, and keep roots from becoming too wet.

DIRECT SEEDING VS. TRANSPLANTING

Deciding whether to transplant or direct seed a crop is based on a number of factors: soil type, climate, the particular crop, weed pressure, available propagation space, and time of year. While most crops can be direct seeded, in an urban farm context

● *Propagation house at Foxglove Farm, Salt Spring Island, British Columbia*

where space is limited and climates can be marginal, growing out transplants is preferable. Transplanting allows for more accurate and reliable results, gives plants a significant head start, reduces weed pressure, and is a more efficient use of seed. It is also much easier to be precise with irrigation in watering transplants. Transplanting is far more time- and space-efficient as a succession crop can already be growing and developing while the field space where it will be planted is still finishing up. On the other hand, direct seeding is excellent for crops that you want to grow at high density, for short-season plantings, and for crops that don't transplant well, such as many root crops.

For all our direct seeding, we use a four-row pinpoint seeder. We've found that it is an easy tool both to use and to teach people how to use, the latter being an important piece of what we do at Sole Food. We've also found that the parts are easy to replace and not too expensive and that we can use the seeder for almost every crop that we grow. After the beds are prepped and amended, we use our planting chart to determine what needs seeding on any given week. The planting chart also specifies spacing and any other details for the time of seeding.

● *Transplanting tomatoes at our farm on Hastings Street*

For transplanting, we use a landscape rake with extensions to mark rows. The extensions, which can be purchased or made from pieces of irrigation pipe, can be spaced to specific crop spacing specifications and will create lines for each row of transplants. To mark in-row spacing, we use simple pieces of lath cut to the desired length.

For maximum efficiency, we have one person marking the beds and two people following and planting. If the marker has a head start, the planters should have no problem keeping up and everyone can move on to the next task together. The first planter drops the plants on the marks created, and the second planter follows and puts those plants into the ground. This is a quick process; we encourage planters to make a hole with one hand, drop the plant in, push soil over to cover the root ball thoroughly, and move on. Having to carry a tool while transplanting slows the process, so box or raised bed production has an advantage as the soil tilth is normally loose enough that planting tools are unnecessary. While transplanting requires good root-to-soil contact without air spaces, it does not require the all-too-common packing soil around plants after they have been placed in the ground. The goal is loose open soil; with a little water and some time, the transplant should settle in nicely.

● *Cutting greens mix*

There are many guides that provide plant spacing information, but spacing is a fairly fluid concept. Spacing affects the size of the final product; the closer plants are transplanted, the more they will compete with each other. Crops planted farther apart not only will get larger but also will tend to mature more evenly. Carrots, for example, will mature a week or two earlier if given more space. Closer plantings create a higher yield, but also result in later and less uniform maturity.

Spacing guidelines in catalogs or on seed packets are good starting points, but you should be willing to test and play to learn what works best for your conditions and needs. At Sole Food, we are constantly experimenting with spacing to achieve the production we want in as little space possible. Be creative: If your experimental carrot planting was too close, thin and sell baby carrots while the full-size carrots mature. Perhaps baby carrots are a crop that can distinguish you from the rest of the market.

BEGINNER'S MIND: WALKING, SEEING, AND RESPONDING

Walking the farm is our most important job. Planning satisfies our need to be organized and to think ahead, but the walks provide real-time, biologically appropriate, moment-by-moment information that informs our decisions. Walking keeps us humble, reminds us of our place in the broader system, and brings us into intimate contact with the real world.

Walk your farm daily, using the same route each time so that you have a baseline, and record the details of what you are seeing. How does each crop look? Is there a response to yesterday's irrigation or cultivation, or the heat or cool of the day? Are the peppers or tomatoes sizing? How do they taste? Is it time to thin beets or harvest carrots? How does the soil appear? Moist? Is there a crust on top? How does it smell? Is it open or locked up? Is there any discoloration in any leaves, any spots, droopiness, curliness, insects, etc.?

● Lissa checking crops

You will come away from these walks with a clear understanding of what is happening with each and every crop, and a detailed list of what needs to be done in response. These walks provide instantaneous feedback; nothing else is more important.

Biological systems never stay the same. We can plot and plan, but in the end nature is always changing. To become a part of the slipstream of our farms, to learn to respond rather than control, to be so flexible that we can bend and stretch and change at any moment—this is the internal challenge of our work. We are not generals in the field always fighting some invading force. Our best work comes when we approach our farms with a beginner's mind: no preconceptions, no fixed ideas or plans, always open to the ever-changing moment.

Walk often, stay open, bring a notebook and a pen.

NEVER WEED, ALWAYS CULTIVATE

Cultivation is the physical act of disturbing the top few inches of soil. It aerates the soil and controls weeds at an early stage before they get established. At Sole Food, our goal is always to cultivate and never weed. And while it is true that there is a subtle difference in how these two terms are used or interpreted, it is an important distinction.

• *Radish and spinach (with primary leaves) interplanted. The radish will finish quickly, leaving the space for the spinach to fill.*

Weeding is what happens when you are too late. Cultivation takes care of any weed pressure often before the weeds are even visible. This is called the thread stage. At this point, the weeds are barely rooted and are vulnerable, particularly on a hot day, to a slight disturbance. The energy required to cultivate is a fraction of what is involved with dealing with weeds after they have had a chance to get a foothold. The timing required to do this well requires constant observation; waiting half a day can mean you are half a day too late.

Choosing the right cultivation tool for the conditions makes the job more efficient, requires less energy, and is more comfortable. There are numerous versions of the traditional hoe; each one has a particular application. Experiment with different tools and find the right one for your body, your soils, local conditions, and particular crop and stage of crop development.

The weed size to look for when determining what to hoe is when the cotyledons, or primary leaves of the weeds, are just emerging. The weeds will almost be invisible unless there are a lot of them. And rather than relying solely on visibility, you may want to base a weeding schedule on predetermined timing. In the summertime, weeds usually germinate on a three- to five-day cycle. After you've cultivated, they're still going to germinate. So every five days or so, you should be cultivating the same bed to keep the weed pressure low. The idea behind timely cultivation is that cultivating a bed can take 20 minutes, whereas hand-weeding a few weeks later could take hours.

There are other strategies besides cultivation with a hoe. A flame weeder consists of a propane tank and a flaming device. The tank can be carried on a backpack, and the flaming device is a handheld torch. The flame weeder is a useful tool when you are direct seeding or when you have a large weed bank built up in your soil. It's especially useful for crops like carrots whose slow germinating time can allow the weeds to get ahead. The idea behind flame weeding is not to burn the plants, but to singe the tiny weeds just before your seeded planting germinates so that your desired crop can get a head start. Note that flame weeding only works on broadleaf weeds, not on grasses.

You can flame a bed after preparation and before seeding, or you can flame it after it has already been seeded. If you wait until after you've seeded, we use a simple test to determine the best time to flame. We put a small sheet of window glass right over one of the seeded areas. Carrots will come up two or three days earlier under that glass; as soon as you see them start to emerge, you should flame the rest of the bed. Otherwise, if you wait until the rest of the carrots emerge, you'll kill them along with the weeds.

With flaming, the goal is to pass that torch across the top of the bed almost as fast as you can walk. You're not trying to burn anything; you are merely trying to explode the cells in the little plants. If you do it right, you will get a crop of carrots coming up into a perfectly clean environment, which is a wonderful thing.

Weeds or Plants out of Place?

- Is a farm really just straight rows of single species?
- The "weeds" we cultivate or remove from our farms are often more nutritious than the plants we leave behind to harvest.
- "Weeds" are the windows into our land. Learn to read them. They often appear to fill an ecological gap, to heal and to balance our soils.
- "Weeds" act as living mulch, and nature's cover crop protecting soils from wind, sun, and water.

WATER WISDOM

When I farmed in California during the drought of the mid-1980s, I received a great education about water. The local water district put forward restrictions based on historical use. Because our farm had already been conserving water prior to the drought, we were given a smaller allotment, while those who had previously been sloshing it around were rewarded. This severe restriction forced me to learn what and how I could grow with very little water.

We adjusted our crop mix in order to focus less on thirsty crops. We researched varieties that might lend themselves to drier conditions, and we changed our planting and growing systems accordingly. Tomatoes were planted deeply and farmed without any water, resulting in fruit that was intensely flavorful. We did the same for melons and beans. We obtained vast amounts of

● *Donna pre-irrigating seed beds*

spent horse stable bedding that was being dumped in the local landfill and used it as mulch, improved our water application methods by moving exclusively to low-volume drip, reduced the frequency of our watering, and became ever more precise with our irrigation timing.

As our soils improved and their organic matter content increased, so did their ability to hold moisture. Poor soils can act like sieves or, conversely, can become so locked up they do not drain. We wanted soils like sponges: absorbing, holding, and gradually releasing.

Open fields have a giant reservoir of moisture that buffers dramatic wet and dry cycles. Containers tend to dry out faster and can have soggy bottoms. As a result, irrigating in containers generally requires more frequency and more careful monitoring. Irrigation also has the added benefit of functioning as a cooling agent, as raised boxes can overheat in warm weather.

• *Spring greens in our original wood growing containers*

No matter where in the world you choose to farm, if you are growing fruits or vegetables, chances are you will probably have to irrigate. Regional climate, soil conditions, and water availability will influence this practice, but there are some universal truths about water and its use:

- While hydrologic systems do replenish themselves, water is a finite and sacred resource and should be treated as such.
- Seventy percent of the world's fresh water goes to agriculture. However, often only a fraction of this makes it to the intended plants or animals due to inefficient transport and application methods.

- Not all water is created equal. Water quality varies dramatically and must be considered when irrigating. Mineral and salt content, temperature, and pollutants all have an impact on soils and on plants and animals.
- Creative irrigation methods can be used to dramatically affect food flavor, texture, and size.
- Application methods and technologies such as drip tape, soil improvements, and mulching can reduce water use, improve food quality, and diminish weed pressure.

Most people think irrigation is about watering plants. Actually, it's more than that. How you irrigate has a profound effect not only on plant growth but also on the flavor of the crop. Wine grape growers know well that by holding back water, sugars and flavors intensify. This can be done even when circumstances don't require it.

On urban farms, water will be expensive, and the source of that water often will be small-diameter, lower-pressure domestic delivery systems not intended for irrigating crops. This requires designing an irrigation system to address these constraints. Drip systems will always be more efficient than sprinklers or flood systems, and are precise in placement of water (reducing weed pressure), low in materials costs, modular, and easy to install, use, and repair. Micro-sprinklers also work well in low-volume and low-pressure systems, and are excellent for germinating densely planted direct-seeded areas and for keeping greens cool and moist in the heat of the summer.

At Sole Food, we have huge disparities in water pressure from site to site. Drip tape allows us to use one system for all of our sites with comparable results. You can bury drip tape and sub-irrigate with it. This is a good strategy for long-term perennial crops like asparagus, and also helps prevent crows, ravens, rats, and coyotes from destroying the tape in order to get water.

Designing a drip system is fairly simple. Drip tape comes pre-regulated with various flow rates. Even so, selecting the right size manifold and installing a pressure regulator helps maintain constant pressure so that each plant is getting the same output of water. This also prevents pressure surges blowing out lines. Investing in an inexpensive filter can help keep fine particulates out of your system so that emitters do not clog, saving maintenance time.

How you irrigate, when you irrigate, and how long you irrigate all affect plant growth and food quality. For example, if you let salad greens go dry, they'll probably survive, but after harvest, you'll notice a bitter flavor, tough texture, and reduced shelf life.

Generally, the goal with irrigation in open fields is to water deeply and infrequently. But at Sole Food, we have very light soils that dry out quickly. In our case, and with any newly planted and shallow-rooted crops, more frequent, lighter irrigations are often more suitable.

Irrigation requires watching and waiting and constantly feeling the soil. My experience has been that people who irrigate on schedules don't always do so well. A schedule is a good baseline and starting point, but there are so many elements at work. Is it windy? How have the plants changed over the past week? Did it rain? How much? And so, like the farm plan, you can create an irrigation schedule, but remember that it's just a rough guide. Your observational skills are the best way to determine when it is appropriate to irrigate.

Different crops require different approaches to watering. I like to water greens the day before harvest to really bring the water content and crispness up. With many fruiting crops like tomatoes, peppers, or melons, holding back on water increases sugars, so we often stop irrigating weeks or even longer before the harvest cycle begins. Often these crops are irrigated immediately after harvest to allow for maximum time without water until the next harvest. There are some fruiting crops, like cucumbers, where watering needs to be more frequent and consistent in order to maintain crunchiness and avoid bitterness. Roots are better harvested when soils are moist but not wet. Crops like greens do not respond well to wide swings in dry and wet—they like to be consistently wet and cool—while other crops like tomatoes respond well to water stress and more dramatic cycles of wet and dry.

Irrigation can also be used to influence the atmosphere and the temperature around the plants. Occasionally we'll have a summer week in which temperatures get above 90°F/30°C. On our sites, which are largely asphalt, it can be significantly hotter than that. At these times, we use micro-sprinklers to cool plants down. Keeping temperature fluctuations to a minimum reduces the stress on plants and can keep them from bolting.

Irrigation is a balance. You don't want to water every time you see a leaf wilting; in some plants, wilting is a natural response to heat, and not necessarily a call to water. For example, if it gets above a certain temperature, squash plants will wilt, even if they are holding excess water. Excess water can also cause wilting by causing roots to rot, so you have to be careful that you're not overwatering. Plants are very resilient, and some plants, having suffered a bit, can produce fruit that tastes better.

PESTS AND DISEASES

It still surprises me that when folks find out that I am an organic farmer the first question they ask is, "What do you do to control the pests?" as if all farming is some sort of heavily defensive act. My answer is always the same, and it is almost always received with some level of doubt and suspicion: I remind folks that pests and diseases are often the result of a system out of balance. It's no different for you or for me. If I push myself hard, don't sleep, and don't eat well, I can hold up for a while, but eventually I will get sick. Just like us, plants require a well-balanced environment to stay well. Plants under extreme stress attract pests and diseases.

The best pest and disease strategy is to create the conditions for dynamic plant health. This starts by nurturing soils so that they are well balanced, nutritionally and biologically, and by providing consistent and stable growing conditions.

There are exceptions, and although there is an arsenal of biological and botanical materials available for the organic grower, at Sole Food, we rarely spray anything unless the problem has reached a significant economic threshold. If the crop is of a value that if lost will seriously affect the farm's ability to pay its bills, and if the pest or disease pressure is at a level that it will ruin most of that crop, only then will we interfere. Otherwise the interference, the cost of the biological or botanical control material, and the impact on the ecological system of the farm are not worth it, and we will let that crop go.

When we do have to implement any pest or disease controls, we try to choose minimally invasive strategies. Exclusion is often our first line of attack. For example, if you don't want carrots with rust fly damage, you've got to keep the rust fly from laying their eggs in the carrots. That's exclusion. How well you do the exclusion impacts the thoroughness of the control, and can also impact the quality of the product. If you're using row cover on carrots to keep the rust fly out, you're going to have tall tops with short carrots because they are not getting enough light. Better to use an insect mesh that allows for full spectrum light transference, but acts like window screen in keeping the flies from entering the planting. At Sole Food, our raised boxes help prevent rust fly issues because the insects don't like flying above a certain height.

Another example of exclusion happens on our fruit trees. Ants are a big issue, not in and of themselves but because of their relationship with aphids. The ants carry the aphids into the trees, where the aphids feed on leaves and leave exudates that inhibit photosynthesis. To combat this, we focus the exclusion on the ants, rather than the aphids. We place a protective material around the tree trunk, usually an old produce label or some duct tape put on backwards. On top of this material, we put a sticky substance called Tanglefoot that the ants cannot cross. Eventually they will form a bridge with their bodies, so the Tanglefoot requires maintenance, often simple stirrng it in order to keep the ants out of the trees.

Other forms of exclusion are tunnel houses and fencing. If you are faced with a pest problem where exclusion won't or hasn't worked, there are a number of botanical and biological controls that can be used. For example, we had a terrible infestation of spider mites in our strawberries. Persimilis, a predatory mite that feeds on spider mites, proved very effective. Bacterial controls such as Bt (*bacillus thuringensis*) are effective as a spray on various caterpillars.

That said, spraying crops even with substances that are relatively harmless and acceptable for the organic grower can have a profound effect on the system in ways that are too often not seen, upsetting the very subtle ecological balance that exists in less disturbed systems. Before you take action, ask yourself why a certain pest issue has occurred. What is going on that would attract this problem? Constant observation and note-taking is useful in monitoring problems as they occur, so you catch things early before any significant damage has taken place. Trapping can be a monitoring method and a really excellent way to know what's happening. For example, to defend against the carrot rust fly, you probably don't need covers over the crop all the time. With traps containing pheromone attractant caps, we can actually monitor when the carrot rust fly is at its peak flight, and keep the carrots covered only at those times.

Another low-cost defensive technique involves creating environments that are inhospitable to pests and/or hospitable to pest predators. For example, slugs like moist environments with lots of organic matter and do not enjoy moving across bare soil. So we don't use mulch as it creates the ideal environment for slugs. We plant potatoes deeply, which prevents flea beetle.

Growing things in their preferred season is another method of pest control, and can teach us a lot about the relationship between climate and plant stress and insects and diseases. For example, kale is a cool-weather crop that grows beautifully and problem-free in the spring, but as soon as the weather warms and dries, becomes infested with aphids. When the cool fall weather comes, the aphids magically disappear, and those same ratty infested kale plants are beautiful again.

Finally, disease pressure can be a bigger issue than pests. In our wet Vancouver climate, we see diseases like powdery and downy mildews that are brought on by lack of airflow and high humidity. Most disease issues are related to specific conditions—high humidity, rain, lack of airflow—so if you can manipulate your environment to reduce those conditions, disease pressure is also reduced. For example, in tunnel houses or enclosed spaces, simple fans can increase airflow and prevent all kinds of potential disease issues.

SEASON EXTENSION

As I have mentioned, having products at the market either early or late is valuable, both in terms of economics and in the relationships and loyalty it forges with customers.

Tools and techniques for season extension allow you to increase the duration that you have saleable fresh products. The most obvious forms of season extension are hoop houses, low tunnels, and row covers. These simple physical devices provide just enough protection so that you can plant and harvest both earlier and later than unprotected conditions would normally

● *Low tunnels*

● *Kale, spinach, and strawberries thrive unprotected while more sensitive crops are under cover*

allow. In almost every climate zone, some form of physical protection is essential for extending market production. This is especially true in the city where space is limited and value per acre must be maximized.

Crop selection can also play a role in season extension; think of selecting crops that match your conditions. At Sole Food, in the shoulder seasons, we stop growing arugula and baby lettuces and substitute baby kale and spinach and a wide variety of chicories—radicchio, escarole, endive, and puntarelle. These greens can be eaten raw or cooked, but taste better in cooler conditions and are more cold-hardy than lettuces. Most of the brassica family is also very comfortable in the cool season and will overwinter well.

Cold storage is also a form of season extension. We don't grow many storage crops at Sole Food because of space considerations and financial return, but potatoes, carrots, celeriac, turnips, and parsnips will all store well for long periods of time through the winter.

Sole Food is often one of the earliest producers in our region due to the warmth and drainage created by growing in raised boxes on pavement. But there are also cultural tricks that you can use for early-season production. Placing a sheet of clear plastic on the top of a newly seeded bed right after seeding traps radiation and warms up the soil. It also keeps the rain off the soil and keeps the seeds from washing out.

Boxes and raised beds are ideal in a climate like ours where conditions are wet and cool because they drain better and warm faster.

Spun polyester or floating row cover adds extra protection, buffers the wind, and allows the soil to heat up a bit more during the day. Floating row cover is effective under many different conditions, not just for season extension but also as an exclusionary system for insect control.

● *Jason pruning and tying tomato plants in one of our high tunnel houses*

Low tunnels—simple wire hoops or bent ½-inch electrical conduit covered with greenhouse plastic—are an inexpensive and modular option for season extension. However, because the air space in a low tunnel is reduced, it will get hotter, which requires more attention and active ventilation.

Less common in the urban context are unheated steel-frame high tunnels. Although they can be moved, these are more permanent structures that are often difficult to rationalize in a city, where you may have to move farm sites from time to time. At Sole Food, we have four 20- by 200-foot high tunnels with roll-up sides and a double layer of plastic that allows us to control ventilation by raising or lowering the sides. This type of season extension, with its more permanent framework and double-layered plastic that provides more insulation, allows us to push some crops much further into the winter and early spring.

Finally, heated spaces are an option, but the operational and energy costs can be extreme, so you must consider the value of the crop relative to the cost of the infrastructure and the energy to heat it. Hot-water tubes that run just below the bed surface are the lowest-energy choice. Forced air or high-efficiency infrared heaters work well, and can generally heat the entire space of a tunnel house, but the cost of natural gas or propane to fuel these systems can be prohibitive.

BIRTH, DELIVERY, AND POSTNATAL CARE (HARVEST/POST HARVEST)

It is too often assumed that the skill in growing food well includes every stage leading up to harvest, and that harvest and what comes after is somehow self-evident and lacking in subtlety or detail.

I'm not sure about other people's farms, but I have yet to have a spontaneous harvest experience where I show up on a Friday or Tuesday harvest morning and find that all the lettuce has cut and washed itself perfectly and fallen into the boxes, or that ripe red or orange peppers have hopped off of the plants and lined themselves up in crates, or that all the ready-to-eat melons magically rolled out of the field and into the truck. And I do not know of any single crop that does not have a particular, variety-specific set of standards, subtleties, and particularities that you must understand if you are to harvest at just the right stage of perfect texture, flavor, and cosmetic quality.

I believe the time just before harvest is the stage in the process that requires the most subtle, careful, and watchful attention to get right. For example, is a strawberry red when ripe, or is it actually crimson in color? What is true for an Albion is not true for a Mara DuBois. Every variety's optimal picking stage is subtly

● *Early-morning carrot harvest*

different. A good strawberry picker has, by tasting and looking, developed a special synapse between eye, palate, and hand that informs a million little harvest choices as he or she travels through a field, moving leaves aside, lifting and looking at the underside of every fruit, and making instantaneous decisions on which one stays behind and which one goes into the box.

Some crops have more subtleties than others. The thing about melons is that while growing them well requires some refined skill (especially in our northwestern climate), harvesting requires almost psychic abilities to get it right. French melons are picked not when they "slip" (fall off the vine) like a Western cantaloupe; instead, pickers must rely primarily on smell, with the support of some key visual cues. During melon season, it is quite entertaining for visitors to witness us crawling around with our noses to the ground, like dogs or pigs sniffing out truffles.

Watermelons, on the other hand, never have any fragrance. We check the underside of the melon that has been sitting on the ground for a change of color from green to white or pale yellow, and look for the drying out of the primary leaf and the tendril closest to the melon, but it is the tapping and listening for that perfect sound that guides our decision. No one seems to entirely agree on approach here, and so I suspect that in the end it is more feeling than anything physical that moves us in our decision to harvest or not.

Melons are especially tricky. At Sole Food, we have always offered a "melon back" guarantee. I sleep better knowing that if people spend five or six dollars for one of our French melons or eight or nine bucks for a watermelon and do not have a religious experience when they eat it, they can get a replacement. We do not require anyone to bring us the already hacked-up remains; we just take their word for it.

● *Cherriette radishes just harvested*

Knowing the particularities of each crop is important in creating the sequence of a harvest. Factors like sales outlets, weather, and labor can affect harvest sequencing, but the top priority is always how to achieve the best quality and flavor and longest shelf life.

On a harvest day, any crop where the leaf is the primary consumable part of the plant comes out of the field first—salad greens, leafy greens, spinach, chard, kale, collards, and herbs. Next might be roots: onions, beets, carrots, and radishes. These can take a long time to bunch, and their leaves can wilt in the sun as well. If you have limited time on your harvest day, you may consider harvesting them a day in advance because they store well in a cooler. Things like potatoes and winter squash can also store well and be done in advance. Peas, squash, and beans might come next on the list. They are not as sensitive to sun and will last better if they are dry when you harvest them. Tomatoes, strawberries, melons, or peppers are always last as they can be harvested in the heat and can benefit from a longer ripening period and a shorter time between harvest and sales.

The harvesting cycle and the volume of harvest is dependent on existing retail or restaurant orders and a knowledge of the demands and potential of each farmers' market or other sales outlet. A Saturday market will almost always generate more sales than a Tuesday market. The time of year, higher-demand holidays, weather conditions that could negatively impact food left on plants, and the personality of each market also determine how much we harvest. We also harvest projecting forward to the next time we'll harvest that crop. For example, if at one harvest some berries are right on the edge of ripeness, they can likely wait until the next harvest, probably only a couple days later, and are better left on the plant to achieve maximum flavor.

When I was farming in California, differences among our market customers also impacted our harvest. For example, eggplants never went to a Saturday or Tuesday Santa Barbara market but were instead hauled on Wednesday to Santa Monica because of the large Middle Eastern population that bought from us there. In Vancouver, we've found that some products sell better to certain restaurants or at specific markets, and we plan our harvest accordingly.

The order of harvest typically reverses as soon as there are frosty mornings (that's in November for us). Some root crops can come out in the morning, but crops like lettuce or kale may have to wait for the afternoon so they can thaw out before harvesting.

Harvesting can consume a disproportionate amount of farm labor time, and so doing it efficiently is essential. For example, if you have 100 feet of head lettuce to harvest, what is the most efficient way to get that lettuce out of the field? Where do you place the box? How do you keep track of how much you have harvested? How do you keep the product from heating up and wilting during the harvest? Ideally, you'll harvest lettuce prior to the sun hitting it to maintain the leaves' turgid quality and to keep them from becoming bitter. If you are working on your own, cut the heads and place them butt-end up in a row in the field. When all have been cut, return with the box and pack all the heads in the box. Always put the same quantity in each box so you know what your total count is; this will aid in packing and distribution. If there are two people, one person does the cutting and places the heads in a row, and the other person boxes them. Full boxes go immediately into the shade and then are hydro-cooled (flooded with water) before going into the cooler or into the shade of the packing area.

When harvesting bunches, the easiest way to maintain a count on what you have harvested is to count out the number of rubber bands/elastics or twist ties equal to the number of bunches required; you will know the task is complete when all the ties or bands have been used up.

When you prepare for a harvest day, leave for the field with everything you need: knives, elastics, twist ties, boxes, and your harvest list (which includes restaurant and/or wholesale orders, or specific upcoming farmers' market sales projections). Always be clear on the order in which you'll harvest, whether it's on the list or in your head. Your goal is to eliminate wasted time, so you can get produce out of the field and out of the heat as soon as possible. You've spent months growing a crop; you want this final stage to be done well so as to preserve the freshness and quality you've built up.

Harvest can also be an opportunity to record yield data and the location of harvest so that your product is traceable. If you place a clipboard and a scale in your packing area, the harvest log (your record for everything that comes out of the fields) can be updated every time produce is dropped off. Note from where in the field it was harvested, how many linear or square feet or growing boxes were harvested, and what the yield was. This data can also be recorded immediately after harvesting as produce is moved from the packing shed into a cooler or other storage area.

We use a year's worth of harvest logs to get a combined picture of total yield per crop (as well as how much of each crop may

• Jordan and Chef David Gunawan checking Sole Food produce delivery

not be saleable). For a sample harvest log, visit: solefoodfarms.com/farming-the-city/resources (use the password Farmingthe city_2018).

It is helpful to have a farmers' market take sheet that lives in a binder in the market truck, to be filled out as the truck is loaded for the market, and when unloading after a market. Tracking this information helps to guide and improve future direct marketing efforts. Find a sample take sheet at: solefoodfarms.com/farming-the-city/resources (password Farmingthecity_2018).

ALIVE AND STILL MOVING

Months of planning, preparation, and hard work eventually result in beautiful food harvested into boxes, bins, or bags. But rest does not come quite yet. The food must be washed and carefully packed, cooled, stored, and prepared for its final journey from the field to the plate.

What you do in those final moments after harvest determines whether all the life, flavor, and goodness in that food will actually be transferred to those who are consuming it.

We often view ourselves like medical technicians, caring for living organisms and transporting them to their destination, just like organs for a transplant. The conditions must be right to keep food that has been pulled from the ground or severed from its mother plant alive and fresh and beautiful. It is essential to know how each of those foods likes to be treated to maximize and preserve its vitality. This includes optimal storage temperature and humidity, the right box for packing, depth of

● *Kelsey sorting and packing tomatoes for market*

in-box stacking, and post-harvest lifespan. Because of the different needs of each outlet, each of these decisions and practices are informed by whether you are direct marketing to the public or to restaurants or selling wholesale to distributors or stores.

Once your produce is out of the field, hydrocooled, and/or placed in the shade or under shelter, it's ready for processing and packing. Some crops, like strawberries, tomatoes, and peppers, won't need to be washed (and in fact will deteriorate if they get wet). Leafy greens need to be washed and sorted, spun or drained, and packed into a box that will allow for the right balance of airflow and moisture retention. Likewise, roots need to be washed and packed in the right box that maintains quality and considers volume and weight for ease of movement. You will need containers with drainage for both leafy greens and roots, but you must also protect those crops from getting dirty again and from drying out in a cooler.

Using plastic packing containers that are stackable allows for easy storage both in a cooler and on a truck. Reusable packing containers are easy to clean and environmentally friendly. Ideally, packing containers should be different than the ones used for harvest containers, to prevent contamination from on-farm sources such as compost and amendments. Good labeling or designated boxes for different crops can help staff differentiate between items without having to open boxes. Packing products with uniform weights and counts simplifies distribution. Case uniformity is also desirable for wholesale accounts used to getting standard cases from distributors.

If you are packing piece orders for restaurant accounts, organizing your labels in advance and generating packing slips can help reduce human error. You may want to consider having a dedicated staff member who packs for restaurants, as the orders can vary widely and include any number of special requests.

Once you have everything packed, get crops into the cooler as fast as possible or make sure they are stored in conditions that will prolong shelf life. Quickly removing field heat from products extends their life span, helping you get the product to the consumer in the best possible condition.

ATTRA, a sustainable agriculture program developed and managed by the National Center for Appropriate Technology (NCAT), has published a helpful document entitled "Postharvest Handling of Fruits and Vegetables." It's available on their website (attra.ncat.org).

CHAPTER 3: MARKETING

WHERE AND TO WHOM

A farm will reflect the personality of its farmer and the surrounding community. The marketing approach should do the same: Location, ethnicity and cultural background of potential customers, the scale you are operating on, the types of products you are producing, and your own personality will determine how and where you market your products.

For example, to do farmers' markets well, you must have an outgoing personality, and be able to create energy in and around your stand. Quality products are foundational, but how you display them and how you present yourself are equally important. You want the stories you tell and the energy you bring to draw people to you and to your food.

While many cities may have multiple farmers' markets each week, they will not necessarily provide enough customers to absorb all of the fresh products you are producing. If you are comfortable with providing significant one-on-one customer service, and you are producing at a high-quality level, restaurant sales might be a good fit. If community building is important to you, a CSA may be a good choice. Many successful farmers combine different marketing options in order to create a more stable support system, fulfill financial needs, and create a business that is personally and socially rewarding.

● *Sole Food's Wednesday farmers' market stand*   ● *Our portable street sign farmers' market branding*

FARMERS' MARKETS

Farmers' markets have made a huge resurgence. Most urban areas now have at least one market per week, if not several. Farmers' market shoppers are a varied bunch: some come to socialize and to pick up a few items, while others use the market to do most of their weekly food shopping.

The most popular market days are Saturdays and Sundays, although midweek markets exist. While they're not usually as busy as weekend markets, they allow for a more even distribution

● *Pok Choi at the market*

of income over the week, and they spread out the harvest, reducing storage needs and allowing for farmers to bring the freshest food to market.

If you are considering doing a farmers' market, it is important to evaluate all of the costs and time commitments. These include market sales labor, travel to and from the market, market fees, setup and breakdown time, and cost of production specific to each market. Keeping good sales and inventory records for each market throughout the year will help you evaluate which markets are most profitable (or whether they are profitable at all) and will guide you in what to grow, and how much, throughout the season.

Continuity at markets builds customer loyalty, so choosing to do a particular market is a commitment that must be weighed carefully. What does it cost to do a particular market, and how much income does it generate? If it's an early-season market and you don't have much product, what are the other benefits of attending that market? Will canceling affect your seniority at the market? How will your absence affect your customer loyalty? How much more valuable are the needs on the farm?

Techniques and Principles for Farmers' Markets

- Design your market display in advance on paper so, when you get to the market, setup is quick and efficient.

- Always be clear about the top several items you need to sell in volume, and provide more space for them, place them in a more prominent location, or otherwise feature them.

- Before the market opens, have a last-minute market staff meeting in which you go over strategies for the day.

- Pile it high and watch it fly: Even if you don't have a lot of product, you can still create an appearance of abundance. Many large food retailers will require employees to attend produce display courses to learn how to do this. Those displays in the produce departments of supermarkets are not giant displays full of produce; there are backstops and steps and blocks behind that produce that provide the illusion of abundance without having to sacrifice perishable product.

- Sample, sample, and sample some more: There is nothing more important than this. We always have one or two full-time people sampling in front of our stand. There are markets when we are handing out $300–$500 worth of free food in the form of samples. Over the years, I've been asked, "How can you give away so much food?" How can we not? Those samples multiply our sales exponentially. For example, handing out Sungold tomatoes, when they are dead ripe globes of pure sunlight and energy and explosive in flavor, sells them better than any sales pitch could. You've spent so many months growing this food—to simply throw boxes on the table and sit down behind them would be a waste.

- Believe in your food. Don't spend the time doing the growing if you don't care enough to complete the cycle wholeheartedly. If you care enough about what you grew and it's good, you want people to taste it. And if it's not good enough, then you shouldn't be bringing it to the market. No one should buy food out of some belief system, because it's organic or local, or out of some sense of charity, or, in the case of Sole Food, because they feel we're doing good social work.

People should buy your food because it's good food, and if not, they should go somewhere else.

- Make sure your space appeals to all the senses, with visually dynamic displays, delicious samples, and fragrance from your products (garlic, herbs, etc.).
- Every product on display should delineate itself visually, and every individual bunch or head or bag should be easy to identify from the whole. This can be done by alternating contrasting colors (not all green in one block) or by alternating the placement of each product (butt end of lettuce alternating with leaf end).
- Bring more food than you think you can sell (it usually does not sell well if left in the field or in the cooler at the farm).
- Be dependable: always have a body of "signature" products available throughout the season.
- Provide a "no questions asked" guarantee for any product you sell.
- Honor loyalty with discounts to regular customers.
- Always have a recognizable likable face and outgoing personality at the market.
- Be outrageous, theatrical, and have fun while selling; you will move more food.
- Never sit down; always be upright and moving, restacking displays, managing product to maintain a visual sense of abundance, and staying face-to-face with customers.
- Don't be in a hurry to disperse the crowds—the more the merrier. A line in front of your booth is good, as long as people are entertained and not waiting too long.
- Use freebies: if you have an item that is difficult to sample, give one to a customer for no charge. They will return for more.
- Reward higher-volume purchases with discounted pricing.
- Educate your customers about less-common products by vocalizing in a few words how to prepare them. Everyone else within earshot will get interested in something they might normally not have picked up.
- Each market will be different: create strategies to accommodate changing products, changing pricing, weather, and season.

● *Pepper display*

ONSITE RETAIL

Another direct-to-consumer marketing option is onsite retail; this could be a produce stand or other setup on or near your farm. If you're located close to a major road or an area where people regularly pass through, this can be a very good sales venue. You will have total control over the marketing. You'll still want to use most of the marketing principles above, but your window for sales can be longer, and you can keep products in the cooler or harvest more as you sell. Onsite retail also provides an opportunity to show off your farm's beauty and bounty to best advantage, and for you to develop a deeper relationship with your community as you give them an opportunity to actually see how the food is produced.

CSA

The CSA model, in its original purity, was conceived as a form of social agriculture whereby community members came together with the farm to provide financial support in advance of the growing season, and then received a weekly share of the farm's harvest—be it a good year or a bad year—throughout the season.

Early on in the CSA movement, which had its roots in Europe and Japan as early as the 1960s, there was much more social and community involvement; members helped with harvests and other farm work and participated in the social structure of the farm. For the most part, what has evolved over the years is more of a different economic and distribution model of food distribution with less community involvement.

When I started a CSA in California in the early 1980s, I sent a letter to our customers saying we were going to close our produce stand and funnel all the food through a CSA. People were very upset. They

● *Winter farmers' market display*

wanted choice—they didn't want the farm to choose for them. I came to realize that just as there are various marketing options that fit the personality, scale, and needs of farmers, similarly there are various options for obtaining fresh food that fit individual lifestyles, personalities, and families. Most folks do not want to be dictated to. So our approach changed, and while we did start the CSA (one of the first on the West Coast), we also continued with a full range of marketing options—wholesale for our tree fruit crops and farmers' markets, on-farm retail, and restaurant sales.

Customers could just date us, or they could cozy up in the relationship and marry us by joining the CSA. We made all kinds of possibilities available, but in the end we discovered that, beyond quality, what was most important was convenience. While many people were becoming more conscious about their food choices and where their food came from, they lived busy lives and wanted to secure their food in the simplest way possible.

● *Cam harvesting yellow carrots*

At Sole Food, we offer two kinds of CSAs: one a traditional farm share, and one we call a market share. Customers can pay up front and have us choose products for them every week during the season (the farm share), or if they want more choice or flexibility, they can buy a market share. It works like a debit system: customers pay either $300 or $500, and receive 15% extra on their share for their loyalty. They are given a gift card that manages their balance, and they can use it all season.

They come to the farmers' market to get their food, and they can choose anything they like without the need to carry cash. If they're away, or if they don't want food in a particular week, their credit remains. At the end of the year, any remaining money on the card rolls over into the next year. Both share types get a weekly newsletter via email so they know what's coming to the market or in their CSA box, what's in season, and any news from the farm.

Customers who join a CSA are typically looking for a more significant connection to you as a farmer, and to your farm. Some people want to be more than just another market shopper; they like supporting local agriculture by paying up front. And you need to return this support: CSA relationships need to be nurtured just like any of the other marketing relationships. You'll be most successful in your CSA program if you can provide a wide variety of high-quality products and if you can connect with your members on a personal level.

RESTAURANT SALES

Like most other sales avenues, selling to restaurants is about building relationships. And these relationships are very different than those you'll build at markets. Selling to restaurants requires building a relationship with a chef or, in some cases, with an owner. Chefs are ultimately the ones who will be using your products, but they can come and go from restaurants and can have widely varied preferences and allegiances, so forging loyalty with a restaurant's owner can be critical as well. And if you create a solid relationship with a chef, they may take you with them wherever they go.

Many restaurants like the idea of moving away from wholesale distribution in favor of the higher-quality and direct relationships they can build with food producers. However, relatively few maintain a real commitment to buying direct from local farms; too often the convenience of one-stop shopping via a single phone call to the local distributor wins out. While the quality of direct-sourced products is always superior, costs can be higher and many restaurants do not want to spend a lot of time on sourcing ingredients.

If you do sell to restaurants, you must have clear sales parameters and protocols. Do you deliver? How many days a week? Is there a delivery minimum? When are orders to come in? How do chefs order? The goal is to strike a balance between offering solid customer service and being realistic about what you can provide. You'll find that it won't make sense for you to deliver for under a certain dollar value, and if your operation is small, it probably won't make sense to deliver every day unless the restaurant is taking a large part of your harvest.

• *Local chefs picking up produce order*

At Sole Food, we deliver twice a week, on Tuesdays and Fridays. These delivery days work well for restaurants that have busy weekends and tend to be closed on Monday. We produce a "fresh sheet"—a produce availability list—that goes out via email on Sunday and Wednesday. It lists the crop available, with variety and description, price per unit, and any specifics about this harvest. For example, maybe one week's kale is cooking quality rather than salad quality. Produce is sold on a first-come, first-served basis, and those with standing orders are given priority. Restaurants have until 5 p.m. on Monday and Thursday to place orders for the following day. This deadline gives us enough time to prepare for what can be a very busy harvest and delivery day.

We give the restaurants a chance to order by email and then follow up with an email, text message, or phone call. These days, phone calls are not always the preferred form of communication, so it's nice to check with your chefs to see what they prefer. Get to know them and communicate often about specials, unique items, or simply to follow up on orders.

How do you get new restaurant clients? Begin by stopping in at their kitchens during non-service hours and dropping off product samples and a business card. Eat in their restaurants occasionally. Look at their menus. Getting to know the menus of the restaurants in your area is the best way to determine whether the relationship would be the right fit. Support them while they're supporting you.

The product that you send to the restaurants should be the best that you have. Sales to restaurants are generally at retail prices, as chefs are given a more time-intensive level of service that requires more communication, packing small, detailed orders, and delivering those orders directly to restaurant kitchens. The expectation of product quality is higher in this relationship, so maintaining consistency and high quality is essential in serving restaurants. Remember, they are paying you for something different than they can get elsewhere.

Consider rewarding restaurants that are consistently ordering larger quantities of a specific item with discounts on that volume. However, before you offer discounts, review your costs of production, marketing, and delivery, and make realistic determinations on your margins and how far you can go. That said, offering discounts on regular bulk orders goes a long way in building loyalty.

The restaurant business is a volatile one. Restaurants come and go and don't always tie up their accounts when they go out of business, so there is some risk in dealing with newer, less-established ones. When you begin a relationship, establish clear payment terms or require COD for some amount of time. Once the client has proven reliable, 30 days are acceptable terms. Extending credit beyond that is very difficult for any small farm. And in order to extend any kind of credit, it's important that you have someone on your team who can follow up, track payments, and help with bookkeeping.

At Sole Food, we provide new and prospective clients with a packet that includes information about our ordering schedule, seasonal availability of our crops, a product and price list, and a credit application. Visit solefoodfarms.com/farming-the-city /resources for a copy of this packet (use the password Farming thecity_2018).

DIRECT TO RETAIL

Direct to retail sales involves selling to the produce manager or owner of a grocery store or supermarket. This normally requires higher-volume production levels and pricing that allows for the markup that retailers must do. The pricing structure for sales to retail will sometimes mirror wholesale pricing values (see next section). Standard markup for produce in retail stores is 100–125%. This markup accounts for shrinkage (product loss) and the overhead of running a grocery. For smaller-scale urban farms, wholesale accounts will be limited. However, if your farm produces a large volume of just a few crops, it could work. You would sell at lower per-pound prices, but you'd simplify your sales process by not having to deal with several different sales channels. One other consideration is that retailers often expect product to be packed in a similar way to what they purchase from wholesale produce brokers and to be harvested at a stage that will hold up longer on the retail display.

WHOLESALE

Wholesale pricing reflects an even greater increase in scale and production to accommodate lower pricing. At the wholesale level, you sell in pallet load volume to a distributor who then consolidates, repacks, and resells that product to retailers and restaurants. The more middlemen between your crop and its end user, the less money you will get. Wholesale pricing can be accessed through reports available online through sources such as the USDA, Rodale Institute, and Canadian Organic Growers that post weekly organic and conventional produce pricing sheets online.

However, if you are not comfortable with direct sales to the public and prefer to focus on growing and selling in large volumes, where a truck arrives at your farm and picks up pallets of produce, wholesale is a good option. This marketing approach is only appropriate for an operation that is producing at very high volume and has the facilities to properly pack, palletize, hydrocool, and load trucks. Because of the scale of most urban operations, this is not a likely scenario for the urban farmer.

VALUE-ADDED

One final type of marketing is producing value-added products— processing your existing crops into new products, like turning strawberries into jelly or jam, cucumbers into pickles, making sauces, freezing fruit into popsicles, drying herbs or fruit, and so on. Value-added products allow you to process cosmetically imperfect or highly perishable fresh food into something less perishable. This results in products you can sell in the off-season, diversifying your income and your appeal.

At Sole Food, we have found that processing our food during the height of the season provides a valuable training opportunity for our staff and eliminates close to 25% of our waste during the season. We also can produce products that customers are excited about, and we have something to sell at the markets during the winter months when there is less fresh food.

Start by researching your local health department's requirements for the specific products you might want to make, and what sales outlets will be options for those products. Doing this in advance will save you a lot of time later on and potentially prevent lost income. When thinking about your line of value-added goods, consider the crops you have large quantities of during the season. Think also about harvest schedules to come up with times you and your crew will have time to process the foods.

If you don't have it on your farm, you will need to source commercial kitchen space nearby; most health departments, stores, and farmers' markets will require that food processing take place in an inspected commercial kitchen. In Vancouver, we are required to provide proof of our processing methods as well as documentation of the type of kitchen we use.

You'll need to create a project budget, and then track expenses as you go along, so you maintain a clear idea of the costs and income of a value-added venture, including what additional value you are or could be deriving from your fresh products.

You've spent time and money to produce what you believe is a quality product, so show it off and attract attention by branding and packaging it appropriately. The quality of the labeling and the packaging are as important as the product that you are creating.

PRICING

The fact that we live in a world that has very distinct haves and have-nots is not an agricultural problem, and it is not the responsibility of farmers alone to resolve. Many of us are doing what we can to address these disparities. However, food security is tied to many factors, such as secure housing and household income. And food can be abundant, but it is not always distributed equally and to all those who need it. There is no agricultural system that can resolve these issues; there are social, environmental, and economic causes that must be dealt with through government policies and by the society as a whole.

In the U.S. and Canada, we pay less for food than anywhere else in the world—10 cents on the dollar, yet we pay for it in less visible ways many times after we've left the checkout counter—in our health, the health of the environment, and in our taxes through huge farm subsidies that go to industrial agriculture. North Americans don't hesitate to buy the nicest car or the nicest television set, but they complain when the best food costs a little more. So I think that farmers have a wonderful opportunity to reeducate people, especially when marketing directly to consumers. Much of the food I have produced over the years has gone to a very narrow segment of society—those who can afford it. Farming is a profession of narrow margins; we have to get the highest return so we can afford to keep farming.

● *Our original farm at the corner of Hawkes and Hastings streets in Vancouver*

At Sole Food, we have set up an interesting paradox. We sell to the top restaurants in Vancouver, and yet we have an employment model that accommodates people who, in many cases, do not have any resources, and could never afford to eat in those same restaurants. Why is the food not going into the low-income neighborhood where many of our staff come from? When we started the Sole Food project, we had to make some clear-eyed decisions. What were our goals? What were the top priorities? We knew that we could not do everything, and we decided early on that providing meaningful employment was our top priority. This is where goal setting is very important, and not just for a social enterprise. When I first came into the project, the ideas were all over the map, but in the end, the priority was the jobs. Ultimately, we have to meet our payroll and we have to train our staff, so we have to sell our food to whoever is going to pay top dollar for it.

When you are conceiving your farm project, think carefully about to whom you want to sell. In the design world, there's a concept called the Iron Triangle. Clients may say about a project, "I want it quick, I want it cheap, and I want it to be quality or well-made." The tough part is that you can't have all three—at most, you can have two out of the three. This same logic applies to farming. You can't grow food cheaply without some essential element of the system suffering, like the soil or your staff or the quality of the food. You're going to have to make some choices—setting prices higher and then offering discounts, for example. Find creative ways to work with your plan and your production costs so you can reach a diversity of people without undercutting yourself.

One of the things that really pushes my buttons is when people underprice their food. Whether they think, "I've got to give them a good deal," or "Somebody else said we should charge this," or "I don't feel like we're worth it"—you've got to get over those ideas. Don't worry about the competition. Do a really good job, and if you are going to charge a little bit more, then stand behind it and don't worry. There is no precise method or science behind pricing the food that you grow. It is a combination of simple local research (knowing what other people are selling similar products for), the relative quality of your offering, knowing all the costs that went into the product, how much volume you have and how quickly you need to move it, and what the market will bear. The goal should always be to have products that stand out, and you should be willing to stand behind them. One should never seek to be the most expensive, but there is no honor in being the cheapest. Underselling your products robs every link in the food chain, from the field to the plate.

YOUR STORY

Telling your story as succinctly and creatively as possible is a critical piece in marketing your farm and your products. People come to farmers' markets and other food outlets for more than just the fresh food. They want to know who you are and why you do what you do; they want to learn something about the land you farm on and how you grow the food they are consuming. Farmers are like a bridge to another world; we fulfill a vital role in society, providing the basic nourishment that people need,

● *Rainbow carrots*

yet we represent only 1.5% of the population. Putting a face to the food we grow, bringing to life our work, providing a link between the city, the land, and the natural world: these are all essential roles that we play. Ultimately our work is about relationships—biological, social, ecological, and nutritional. You can strengthen those relationships by using words and pictures to share your story and to bring eaters into your world. Telling your story enriches the experience of those you serve, improves your sales, and deepens your relationship with your community.

CHAPTER 4:
PEOPLE AND PRINCIPLES

THE HUMAN ELEMENT

When we started Sole Food, we were very deliberate about our intentions. The goal was to provide training and employment to individuals with barriers. This intention and focus on the human element has guided everything we do at Sole Food.

Jobs are the priority, so we want to employ as many people as our farms can support. We focus on hiring people who are facing barriers to traditional employment mostly because of addiction and mental illness, but who can still function in a job. We place less emphasis on resumes, interview etiquette, and experience and more emphasis on intention, need, and interest.

We want our staff to succeed, so we don't set hours that they can't meet. This means we don't create schedules that we force people into simply because it makes more business sense. We work around prior engagements and abilities, and strive to help our employees maintain a balance. Ultimately, our goal is to balance the hard work with helping staff improve their quality of life.

We don't have high turnover. We have extremely high loyalty from our staff, many of whom have been with the farm for two to eight years. This, plus our social mission, means that aside from the few of us who are in management, we don't have a lot of room to hire people who do not have barriers. Farm interns can provide inexpensive labor, but you will have to make the commitment to train and retrain year over year, and you will sacrifice efficiency. Volunteers also take work (and, as a result, money) from those in the neighborhood whom we aim to serve and employ.

Jesus harvesting Hakurei turnips

Our emphasis on employment also means that we generally have very high payroll costs. Having many part-time employees, as we do, is not usually as efficient financially as having fewer employees who are full-time.

At Sole Food, we have anywhere from 20 to 25 employees at any one time. They aren't all full-time, but there are still 25 individuals that we need to support. Management of payroll and human resources can be very challenging.

Ask yourself what your operation requires based on all the decisions you've made up to this point. How do you want to spend your time? If you have a staff of more than five people, chances are you will spend a lot of time managing them. A small organic farm requires, on average, about one person per acre, assuming you have some kind of light machinery support. If you are working very intensively or purely by hand, two or three people per acre might be more appropriate.

At Sole Food, we have a management team of four including an executive director, administrative coordinator, marketing/distribution coordinator, and operations director, and about 20 other staff members.

● *Homemade branding*

Regardless of who and how many your staff are, establishing good communication from the start is essential. Give clear instructions when you're explaining a task. Be honest when you don't know the answer to something. Give feedback on work and suggestions for improvement. Establish regular staff meeting times where you can communicate information about the big picture and receive comments and questions in a more relaxed setting. Giving regular employee evaluations is also important in communicating your interest in staff well-being. It's an opportunity for them to give and receive feedback and for you to gauge employee engagement. Your goal is to make the farm work for everyone, and checking in regularly with your staff will do that.

Much of the value of the work that Sole Food Street Farms and many organizations similar to it do is often not captured solely on a financial statement's bottom line. Fortunately there are now

● *Alain showing Cam how to harvest French Breakfast radishes*

good protocols for assigning dollar values to the social, ecological, and educational work that we do. In 2013, Queens University (Kingston, Ontario) did a full study of our work to demonstrate our broader financial impact on society. They determined that Sole Food's efforts had both economic benefit (we saved taxpayers $247,370 annually) and environmental benefit (our farm, namely our orchard and our CSA program, was responsible for 3,500 kg of CO_2 removal each year). Finally, for every dollar we pay our staff, there is a $2.25 savings to the broader society in individual, social, and environmental benefits. (In late 2018, a researcher from Quest University (British Columbia, Canada) updated this study and found the social return on investment (SROI) was in fact $5.07 for every dollar paid to our staff.) On top of this, we contributed to society in ways that were more difficult to quantify, including improving the well-being of our employees.

General Farming Principles

The following are some of the ideas and basic principles that I have adopted over the last 40-plus years I have been farming. Add your own as your career progresses, and refer to them regularly, using them as simple guides or reminders. Though they were originally created for rural farms, they apply almost equally to an urban setting.

- On a small farm, it's not the big ideas that make things successful; it's all the little things done really well.
- Grow only those crops you or your family like to eat.
- Be early and be late with your crops.
- Maintain a consistent supply of a small body of signature products.
- Be clear about why you want to do this work.
- Design, plant, and harvest your farm on paper before a single seed is ordered.
- The smaller the farm, the higher the value of every leaf, every fruit, and every root.
- Crop diversity can be a form of financial security—it allows you to ride through changing climates, conditions, and markets.
- The more diverse your crop mix, the more skill you require. Start with simple crops and narrow crop diversity, then expand and diversify as you improve your skills.
- Become intimate with the land you will farm: know how the light, air, and water move throughout the year. Learn the unique characteristics of every field, the subtle changes in soil, climate, exposure. Know your land as intimately as you know a lover.
- Understand your land's history: Who came before? What were they doing?
- If you want your kids to farm, don't raise them on one.
- Stay debt-free, or at the least, limit your debt.
- Know your market before you plant the first seed.

● *Our farm on Pacific Boulevard in Vancouver*

URBAN FARMING MANIFESTO

I developed the following manifesto over a 10-year period. Some of the ideas may sound radical; others are terribly obvious. Some are practical, some more ideological. They are all focused on the municipal and on individual ways to address what I consider to be some of the most prominent challenges in how we feed ourselves.

- Every municipality should establish publicly supported agricultural training centers in central and accessible locations. I'm not talking about think tanks or demonstration gardens. I'm talking about working urban farms that model not only the social, cultural, and ecological benefits of farming in the city but the economic benefits as well. We can talk about all of the wonderful reasons to farm in urban areas, but until we can demonstrate that it's possible to make a decent living doing it, it's going to be a tough sell.

- Regular folks are now so removed from the work of farming that they need to literally see what's possible. They need access to those who have maintained this knowledge and those who are serious and active practitioners. Every city should have teams of trained farm advisors in numbers proportionate to the population devoted to urban food production. Those agents should operate out of their local urban agriculture centers to run training workshops and classes; they should also venture out into the community to provide onsite technical support in production, marketing, and food processing and preparation.
- The nutrient cycle that once tied farms with those they supplied has been interrupted. We need a full-cycle food system that allows for the return of organic waste via central regional composting facilities that can support the nutrient needs of both urban farms and farms on the fringe of our urban centers. Every community could be composting all of their cardboard, paper, old clothing, shoes, restaurant and grocery store waste, and

● *Sole Food Farm flanked by bridges and buildings*

so on. We need to reduce what comes into our communities from elsewhere, but we also need to reduce what leaves those communities, especially if it has nutritional or soil conditioning values for our land.

- At one of the open-field farms I work, the fields seem to have as many rocks as grains of soil. Removing those rocks represents a huge amount of work for me, but each one of those rocks also represents an enormous amount of embodied energy, if I could just release it. Every community should own a portable rock grinder that could be taken to farms and used to grind rocks in and around fields that contain essential minerals that are now being mined elsewhere at great ecological cost. There are huge holes in the world, entire mountains removed, to supply minerals such as gypsum and lime and rock phosphate to our farms. Cement plants can produce valuable rock dust as well. Let's put it to use. We cannot talk about a sustainable agriculture unless we address where the minerals—especially phosphorous—are going to come from. We've all heard about peak oil; we now need to prepare for peak water and peak phosphorous. We can grow food without oil, but we cannot grow food without phosphorous and water. Phosphorous is a mined mineral, which now has limited reserves, most of which are located in China, Morocco, and the Western Sahara. Some scientists believe that, at our current rate of use, remaining reserves will be depleted within 50 to 100 years.
- Let's get over our phobia around human waste, stop spending billions of dollars to flush it away and pollute our rivers and oceans, and start recycling it onto our farms and gardens. Urine is the best local source of phosphorous, and we need to figure out creative ways to recycle it.

- Every community should support the construction and funding of a permanent, covered, year-round farmers' market space in a central, well-trafficked location. Providing this type of physical space is just as important to our civic health, if not more, as the public swimming pools, sports fields, schools, churches, and libraries.

- Every new permit for a housing development should be contingent on inclusion of an approved food-production component on a scale relative to the number of people who will live in the development. And every new office or retail building should be engineered for a full-scale rooftop food production component, including greenhouses warmed by the spent heat vented from the building.

- Every neighborhood school and church should be required to restructure existing facilities to accommodate cooperative canning, freezing, and dehydrating services for their neighborhoods during non-peak hours.

- Every real estate transaction should include a small urban farmland preservation tax from which lands could be purchased specifically for the production of food. Those lands would have protective easements that would require agricultural use in perpetuity.

- A great deal of privately owned arable land currently lies fallow. This land could be made available to new farmers under long-term leases. We need to recognize that there is not necessarily any relationship between land ownership and land stewardship. The only requirement for land ownership in our society is access to capital. That's not enough. I believe that ownership of land should come with a set of responsibilities. Building inspections are common practice in many real estate transactions; we should require land inspections, including ecological assessments and baseline documentation, on every piece of land over five acres. Every land purchaser should be required to attend a stewardship and restoration training course based on the particularities of that piece of land. This will move land away from its status as commodity and bring some sense of stewardship into ownership.

- When I was in school, my favorite classes were woodshop, metal shop, mechanics, and home economics (which included cooking and sewing). Those subjects were well respected. I looked forward to shop class far more than math or science or English. It was a time when I could make something real and tangible. (Every woodshop teacher I've known was missing a finger or two, and I made the connection very quickly between those missing fingers and the machines that we worked with.) Life skills classes are coming back into schools, but we need to give farming and cooking and mechanics and plumbing and carpentry the same status and attention as math or science or English.

- It sounds radical, but in the future, full-time professional farmers may no longer have the luxury of raising fruits and vegetables. This should become the responsibility of individuals and families to grow for themselves in their front and backyards, on their balconies and rooftops, and in community garden plots. We could probably survive without another carrot or tomato, but we cannot live without grains and beans and other protein sources. Every municipality should initiate a phase-out of all home lawns—effective immediately—and also provide neighborhood training programs and technical support for home and building owners to replace those lawns with food production.

- It may be that, along with growing food, the real work of farmers in the future will be the sequestration of water and carbon. Anyone who has land, or is managing land, has a huge opportunity and a responsibility to address two of our greatest global challenges—water and climate. Slowing and spreading surface water and allowing it to percolate and not run off, along with learning to use land and improve soils to store and hold carbon, are urgent and essential roles that farmers need to play now and into the future.

AN URBAN ORCHARD

(adapted from *Street Farm: Growing Food, Jobs, and Hope on the Urban Frontier*, Chelsea Green Publishing, 2016)

Establishing an orchard on a derelict lot in the middle of the city has been a challenge. Planting fruit trees in boxes rather than in the ground (for portability, in case we needed to move our farm) was just one more experience of jumping off the cliff without knowing what we would find on the way down. This willingness to take chances, to experiment, has fueled the great agricultural experiment for thousands of years.

When I worry about the potential constricted container space of our urban orchard, I remind myself about one of my basic understandings about growing fruit. Whether it is strawberries, figs, melons, or peaches, the ones that taste the best are almost always those that have suffered. Having everything you need all of the time—living a life where the road is straight and smooth and without challenge—does not develop character and depth. Those misshapen, sometimes smaller strawberries or apricots inevitably have the most sugar, intensity, and complexity of flavor. Is it possible that the trees we are growing, trees that will surely face adversity in constricted containers, in the harsh environment of the city, in a marginal climate, will grow up to make babies that will be so tasty they'll knock our socks off?

The corner of Main and Terminal in downtown Vancouver was about as unlikely a place for an orchard as I could imagine. And yet this site now contains close to 500 trees, thriving and making fruit: persimmons, figs, apples, cherries, plums, pears, and quince. The trees have trunks four and five inches in diameter,

● *Plums and apples in our urban orchard, with an understory of carrots*

● *Shiro Plums on four-year-old trees in Sole Food's orchard*

and herbs—chocolate mint, spearmint, peppermint, lemon mint, savory, lovage, sage, lemon thyme, marjoram, Greek and Italian oreganos, lemon balm, chives, and more—fill the space between the trees, spilling over the sides of their containers.

The orchard feels complete, like a little self-contained dynamic world, the tree branches reaching over the narrow pathways between the rows, quince touching apple, plum and pear and cherry intertwining, fruit-laden fig and persimmon limbs hovering over black asphalt paving. On a summer day, the hum of insect activity is intense; bees and wasps and flies almost drown out the din of cars and bicycles and pedestrians.

If you're walking on the sidewalk along the edges of this space, the allure of fruit and shade and dynamic plant life is real. It's inviting. The temptation to climb the fence can be overwhelming. In 2015, close to 200 pounds of apples disappeared in a single night. The thieves left behind the handmade extension tools they used to harvest the fruit, devices that were impressive

● *Persimmon trees producing heavy crops in the fourth year of the life of our orchard*

in their ingenuity. A couple of years later, we arrived one morning to find that the entire fig crop had been harvested and the leaves stripped from the trees.

One day I arrived at the orchard and found an elderly homeless man harvesting herbs from below the trees. When I asked him what he was doing, he told me this was "God's food," and kept on cutting. I responded that God had a lot of help from our crew in planting and nurturing those herbs and that we needed to sell them in order to provide paychecks and a place to come to learn and work. I told him he could keep what he had harvested, that we are always willing to share if we are asked first, and that he should leave the site.

These events remind me of an ongoing internal conflict. Deep down, and apart from the farm production mentality that often gets in my way, I like the idea of opening up our orchard for free-form, Garden-of-Eden consumption and enjoyment.

I struggle with the idea that our farm or any other farm is cut off, protected, and inaccessible. If you spend your life on the street, and you're hungry, harried, hassled, and tired from the stress of trying to survive, you could let go here, and immerse yourself in the magic and healing and pure edible sweetness of this little oasis. You could allow it to wrap itself around you and infuse you with its aromas, its sweet flavors, its safety and shelter, and its simple beauty.

INDEX

flats, for transplanting, 41
foliar feeding, 26
food production, by individuals, 98
fruiting crops, 55, 66. *See also by name.*
fruit trees, 57, 99. *See also orchards.*
full-cycle food system, 95–96
fundraising principles, 6–7

goals, 4, 26, 31, 39, 48, 80, 86–87, 91
greens, 27, 28–29, 53, 54, 55, 60, 65, 66–67, 69

harvesting, 63–68, 84
harvest plans, 33–34
human waste, recycling of, 96

interns, 90
irrigation, 53–55

kelp solutions, 26

laboratory soil testing, 24–25
land selection, 7–13
land stewardship, 97–98
leasing land, 8–10
lettuce, 60, 66–67, 75
loans, 4–5

management, 31, 90–91
marketing, 71–85
market research, 27, 28, 84, 87
melons, 50, 55, 64, 66
micro-sprinklers, 53

minerals, in soils, 96
mission statements, 1–2, 31, 90
mulch, 50–51, 53, 58
municipal permits, 10–13, 84, 97–98

native soils, 15
nonprofits, 5–6

observation skills, 24, 25, 46–47, 48, 54, 58, 63–64
onsite retail, 76
open-field agriculture, 8, 15, 22, 37, 39, 42, 51, 54, 96
orchards, 99–103

packing containers, 70
pests, 56–58, 61
phosphorous, 96
planting plans, 34–35, 44
pricing, 75, 82–83, 85–87
principles, of farming, 93
proactive management, 31–32
produce processing, 68–70
propagation, 41–42
proposals, 9–10

reactive management, 31
record keeping, 36, 37, 46, 67–68, 73
restaurant sales, 33, 66, 70, 71–72, 79–82, 86
retailers, sales to, 82–83
rock grinders, 96
root crops, 60–61, 65, 66, 69–70. *See also by name.*
row cover, 57, 59–60, 61

ABOUT THE AUTHOR

MICHAEL ABLEMAN is the cofounder and director of Sole Food Street Farms and one of the early visionaries of the urban agriculture movement. Ableman has worked as a commercial organic farmer for the last 45 years. He is the founder of the nonprofit Center for Urban Agriculture, and has created high-profile urban farms in Watts and Goleta, California; and Vancouver, British Columbia. Ableman is the author of *From the Good Earth* (Abrams, 1993), *On Good Land* (Chronicle Books, 1998), *Fields of Plenty* (Chronicle Books, 2005), and *Street Farm: Growing Food, Jobs, and Hope on the Urban Frontier* (Chelsea Green, 2016).

For more information about Michael Ableman go to www.michaelableman.com or www.foxglovefarmbc.com

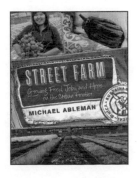

For more information about Sole Food Street Farms, read *Street Farm: Growing Food, Jobs, and Hope on the Urban Frontier* (Chelsea Green, 2016).

More Resources from New Society Publishers

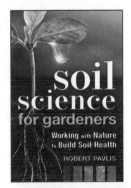

Soil Science for Gardeners:
Working with Nature to Build Soil Health
Robert Pavlis
6 x 9" / 224 pages
US/Can $18.99
ISBN 9780865719309

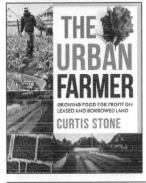

The Urban Farmer:
Growing Food for Profit on Leased
and Borrowed Land
Curtis Stone
7.5 x 9" / 240 pages
US/Can $29.95
ISBN 9780865718012

The Farmer's Office:
Tools, Tips and Templates to Successfully
Manage a Growing Farm Business
Julia Shanks
7.5 x 9" / 288 pages
US/Can $24.95
ISBN 9780865718166

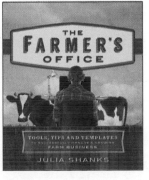

The Market Gardener:
A Successful Grower's Handbook for
Small-Scale Organic Farming
Jean-Martin Fortier
8.5 x 8.5" / 288 pages
US/Can $24.95
ISBN 9780865717657

ABOUT NEW SOCIETY PUBLISHERS

New Society Publishers is an activist, solutions-oriented publisher focused on publishing books for a world of change. Our books offer tips, tools, and insights from leading experts in sustainable building, homesteading, climate change, environment, conscientious commerce, renewable energy, and more—positive solutions for troubled times.

We're proud to hold to the highest environmental and social standards of any publisher in North America. This is why some of our books might cost a little more. We think it's worth it!

- We print all our books in North America, never overseas.

- All our books are printed on **100% post-consumer recycled paper**, processed chlorine-free, with low-VOC vegetable-based inks (since 2002).

- Our corporate structure is an innovative employee shareholder agreement, so we're one-third employee-owned (since 2015).

- We're carbon-neutral (since 2006).

- We're certified as a B Corporation (since 2016).

At New Society Publishers, we care deeply about *what* we publish—but also about *how* we do business.

Download our catalog at https://newsociety.com/Our-Catalog or for a printed copy please email info@newsocietypub.com or call 1-800-567-6772 ext 111.

New Society Publishers
ENVIRONMENTAL BENEFITS STATEMENT

By using 100% post-consumer recycled paper vs virgin paper stock, New Society Publishers saves the following resources:[1] (per every 5,000 copies printed)

14	Trees
1,258	Pounds of Solid Waste
1,384	Gallons of Water
1,805	Kilowatt Hours of Electricity
2,286	Pounds of Greenhouse Gases
10	Pounds of HAPs, VOCs, and AOX Combined
3	Cubic Yards of Landfill Space

[1] Environmental benefits are calculated based on research done by the Environmental Defense Fund and other members of the Paper Task Force who study the environmental impacts of the paper industry.

Certified

B Corporation

MIX
Paper from
responsible sources
FSC® C016245
www.fsc.org

new society
PUBLISHERS
www.newsociety.com